Puzzles and Paradoxes

in
Relativity and
Cosmology

Oliver Linton

joliverlinton@gmail.com

ISBN 978-1-4710-5850-9

9 781471 058509

Contents

Conclusion

Appendix

Introduction

Many, if not most, revolutionary scientific theories have come about because someone observed something strange and began to ask – why is that? Many of these discoveries have been made almost by accident when the discoverer was actually trying to do something else.

In 1644 Evangelista Torricelli was attempting to create a vacuum by using the weight of a column of mercury in a long glass tube. One day he noticed that the height of the column of mercury in the tube varied with the weather and concluded that the reason was that air had weight and that, in fact, we lived 'at the bottom of an ocean of air'. The daily variation in pressure was due to 'waves of air' passing over us on the surface of this ocean. In a sense, this accidental discovery was the start of the whole science of meteorology.

Hans Christian Oersted is said to have noticed, while delivering a lecture on magnetism, that a compass needle deflected when he switched on a nearby current thus initiating the study of electromagnetism. While it is true that he, and others, had been searching for a link between the two phenomena for many years (sailors had long known that ships compasses were unreliable in a thunderstorm, for example), it seems highly probable that the precise arrangement of wires and compass needle on the bench that day was largely accidental.

In 1896 the French physicist Henri Becquerel was investigating the possibility that phosphorescent substances might emit the newly discovered X-rays. The procedure he used was to wrap a prepared photographic plate in black paper; place a phosphorescent compound on the plate and expose it to the sun in order to activate the phosphorescence and then see if any rays had penetrated through to the sensitive plate. One day, having prepared some plates and placed some Uranium salts on top the clouds came over and obscured the sun so he placed the plates into a drawer and forgot about them. Some time later he remembered the plates and decided to develop them anyway, confident that they would be blank because the salts had never been exposed to strong light. Imagine his astonishment when he found that the plates were heavily exposed showing that Uranium salts emit something like X rays all the time! This discovery quickly opened up a completely new field of research – radioactivity.

Of course, not all such discoveries are happy accidents. When Isaac Newton noticed that a glass prism caused the image of the Sun to be elongated and split into colours, he was only refining an observation that the makers of chandeliers had been aware of for centuries. Nor would he have been the first to ask the question why this was so; he was, however, the first to formulate a coherent answer to the question. Darwin and Wallace were able to ask the question why do species exist only because they had both amassed an immense amount of observational experience over many years.

But just occasionally a genius comes up with a new theory, not because he has discovered something new which needs explaining, but through a deep seated belief that *it must be so*. Often, it has to be said, such *a priori* theories prove utterly false. Aristotle's theory of the Four Elements and Kepler's 'Harmony of the Spheres' spring to mind. But Galileo's idea that uniform motion was not a condition which had to be constantly maintained but was itself a state, no different in principle to the state of rest, had absolutely no observational evidence whatsoever in support of it. The famous experiment involving the dropping of weights from the leaning tower of Pisa, if it ever happened, was probably carried out by Galileo's detractors to prove him wrong rather than the reverse and the idea of proving Galileo's ideas about the relative nature of motion by dropping a cannon ball from the mast of a ship is laughable. But what else could he say? He just knew he was right and that, if only you could pump all the air out of the Duomo, a hammer dropped from the top of the dome would fall no faster than a feather and that, if chariots and ships existed which could travel smoothly at high speeds, it would be as easy to pour a glass of wine on board as it is in a tavern in a piazza. Yes, his intuition was not perfect. He thought that uniform motion in a circle was the natural state of things, not motion in a straight line, and that the tides proved his theory that the Earth moved. He was wrong on both counts but Newton put these things right within a few years.

Another *a priori* theory was James Clerk Maxwell's theory of electromagnetism. By 1860 it was well established that a stationary system of electric charges created a static electric field (Coulomb's law), a constant current created a constant magnetic field (Oersted's law) and that a changing magnetic field could create a constant electric field

(Faraday's law). Maxwell realised that these three phenomena positively demanded a fourth – namely, that a changing *electric* field would generate a constant *magnetic* field. Maxwell had no experimental evidence for this phenomenon, he just *knew* it had to be so.

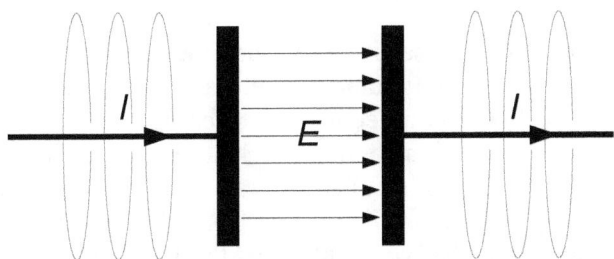

In the above diagram a current I is flowing onto the left hand plate of a capacitor and off the right hand plate. No current flows from the left hand plate to the right hand plate; all that happens here is that the electric field E increases. The currents in the wires obviously create a magnetic field. Does the magnetic field just stop? Or does the changing electric field create a magnetic field also?

This profound insight led to the prediction that light was itself an electromagnetic wave and that other waves with different frequencies and wavelengths would yet be discovered.

Einstein's two great Relativity theories were also based on nothing more than a conviction that *it must be so.*

As an old man Einstein recalled that, as a teenager, he often wondered what it would be like to ride on a light beam; later, when he had learned about Maxwell's theory he became uncomfortable with the idea that you could ever travel beside a light beam and see it, as it were, 'frozen' in space. Maxwell's equations did not seem to require a frame of reference against which to measure the speed of light nor did they seem to need an 'aether' through which the light had to travel. Once you had accepted the idea that electric and magnetic fields could exist by themselves in a complete vacuum, there seemed to be nothing to stop a system of changing electric and magnetic fields propagating themselves through that vacuum at the speed predicted by Maxwell.

But this realisation brought with it an apparent contradiction. How could two observers in relative motion through empty space get the same answer for the speed of light? For most scientists of the day, this riddle had only one possible solution. Light travelled at a fixed speed through absolute space (whether or not an actual aether existed) and that the observer who was at rest with respect to absolute space would measure the correct value for the speed of light and the other observer would get it wrong.

But Einstein was not satisfied. By the time he was 21 years old, he had acquired a teaching diploma in maths and physics (in which subjects he excelled) but failed to get a teaching post, so a couple of years later he took up a position in the Swiss Patent Office in Bern. The duties there obviously did not tax his brain over much and he had time to work on several problems in physics besides the nature and behaviour of light. Indeed, if he had only published the 1905 paper on Brownian motion or the paper on the photoelectric effect and nothing else in his whole life, he would still have earned his place alongside the likes of Lord Kelvin and Max Planck as one of the greatest scientists of the day but it was his paper on Special Relativity that set him apart as a genius. Like Galileo and Maxwell before him, he just *knew* that the speed of light had to be constant for all observers, whatever their relative motion[1]. And with that conviction he set about working out exactly what the consequences of that assumption would be and where the logical contradiction would rear its ugly head. What he discovered was that rulers would shrink and clocks would run slowly – but there was no *logical* contradiction at all. In fact, it all worked out beautifully.

Later that same year a second paper extended his theory to include energy and momentum and it was this second paper which included the now famous equation

$$E = mc^2$$

The Special Theory of Relativity is usually considered to apply only to observers in constant motion. Actually it applies equally well to the dynamics of accelerated motion provided that the concept of a force is defined appropriately. But Einstein was still not completely satisfied. He had shown that no experiment could be performed inside an inertial

---

1   It is true that an experiment had been performed in 1887 by the American physicists Michelson and Morley which provided strong experimental evidence that this was so but, apparently, Einstein was unaware of this result.

laboratory to determine whether or not it was moving – indeed, this principle is essentially the foundation of Special Relativity – but it was still possible, he realised, to determine whether or not the laboratory was *accelerating*. On the other hand, he knew that gravity acted very like accelerated motion so, perhaps, the principle could be extended to include the impossibility of distinguishing acceleration from gravity – an assumption known as the Equivalence Principle.

In the event it took him 10 years of concentrated effort and a lot of new mathematics to prove that there were no logical contradictions in this idea either. The result was the General Theory of Relativity which was published in 1915-16.

At first, scientists were understandably cautious. Special Relativity was elegant but seemed to have little relevance to contemporary physics – after all the fastest thing on the planet at the time was a rifle bullet. Even the recently discovered 'cathode rays' appeared to travel at speeds much less than the speed of light. The only experiment which supported Einstein's theory, the Michelson-Morley experiment in which two light beams are set to race along two arms at right angles, could easily be explained either by 'aether drag' or by a physical contraction of one of the arms. The prediction that all particles had energy by virtue of its mass seemed to have more relevance, but this only became apparent in the subsequent decade with the discovery of the transmutation of the elements and the energy released in nuclear reactions.

By the time the General Theory appeared, the Special Theory had yet to be conclusively demonstrated experimentally (although the majority of scientists accepted it); but the General Theory seemed to offer some categorical predictions which could be verified with precision. And so it was that in 1919 a number of expeditions set out to observe some stars during a total eclipse of the Sun. According to Maxwell's electromagnetic Theory of light, the Sun's gravity should have no effect on the passage of light nearby. On the assumption that light was a stream of particles – a view supported by evidence from the photoelectric effect and also a simple interpretation of the Equivalence Principle – it was expected that the light would be deflected by a certain small angle. But the full General Theory which takes into account the distortions of spacetime produced by the mass of the Sun predicted an angle twice this amount.

When the results of the astronomical observations were analysed and published they were hailed as a complete vindication of Einstein's theory. For some reason the editor of the New York Times decided that he could make a big story out of this and published the famous headline 'LIGHTS ALL ASKEW IN THE HEAVENS – but Nobody Need Worry' and overnight, Einstein and his theories became a household name.

LIGHTS ALL ASKEW
IN THE HEAVENS

**Men of Science More or Less
Agog Over Results of Eclipse
Observations.**

EINSTEIN THEORY TRIUMPHS

**Stars Not Where They Seemed
or Were Calculated to be,
but Nobody Need Worry.**

Public reaction to the theory was, understandably, confused. Most readers would have understood little of what was being claimed but anyone who took the trouble to read further could be forgiven for reacting with sheer disbelief. The very idea that space and time could be mistaken one for another seemed to run counter to 2000 years of human history; the very thought that two events could change their temporal order depending on your point of view seemed totally illogical; and the idea that a wandering space traveller could return many years younger than his twin brother, frankly absurd. Einstein was not just putting forward a new physical theory, he was shaking the very foundations of our beliefs about the universe we lived in.

In 1931 a book was published (in German) titled 'A hundred authors against Einstein'. Einstein's dry comment was 'If I were wrong, one author would be enough.' But the debates have not stopped. Even while convincing experimental evidence for the reality of time dilation from the decay of muons was pouring in, the respected physicist and Fellow of the Royal Astronomical Society Herbert Dingle argued in 1950 that relativity 'unavoidably requires that A works more slowly than B and B more slowly than A – (a proposition) which it requires no super-intelligence to see is impossible.'

The reality of time dilation and the resolution of the twins paradox has largely been accepted now; but length contraction still causes much heated debate in science forums on the internet these days, in particular in respect of the 'Broken Rope' paradox (also called Bell's spaceship paradox) and the 'Rotating Space Station' paradox (also called the Ehrenfest paradox).

General Relativity is usually considered to difficult for mere mortals to understand and attempts to make its effects comprehensible are usually restricted to fanciful accounts of black holes, wormholes and time warps. But in these days of fantastically accurate GPS satellites, the phenomenon of Gravitational Time Dilation is of great importance and it is no more difficult to understand that kinetic time dilation – a process which is not helped by the fact that many books and websites still give the totally false impression that Gravitational Time Dilation is somehow related to the strength of gravity at the point in question.[2]

As for Cosmology, while here are many excellent accounts of our current understanding of the history of our universe, almost no attempt has yet been made to explain to the layman how we can see objects which are so far away that it would take twice the age of the universe for light to reach us from them.

I will not claim to have the last word on these issues but nothing sharpens the mind like a good puzzle and serious consideration of these problems can do nothing but help us understand one of the most remarkable theoretical discoveries ever made by a mortal man.

---

2   For example one website states that 'The stronger the gravity, the more spacetime curves, and the slower time itself proceeds.' A similar error can be found in Martin Gardner's otherwise excellent book *'Relativity Simply Explained'* Dover 1996 p. 116.

Part I: Length and Time

Relative Motion

Arthur and his sister Betty were travelling up to London on the train. Feeling the need to stretch his legs, Arthur wandered off down the train for a walk. A few cars down he came across the refreshment bar and bought himself a coffee. When he returned, coffee cup in hand, his sister was not unreasonably a bit cross.

'You didn't think to get me one too' she complained.

'No. I thought...'

'You didn't think at all. Men are so selfish! Where did you get that coffee anyway?' she said, rising from her seat.

Not feeling particularly charitable towards his sister, Arthur replied 'Just north of Watford, actually.'

'Pah!' expostulated Betty and stomped off in the direction Arthur had come from.

This silly little anecdote makes an important point. When you are in a moving train, there arc two obvious frames of reference which you can use to describe the position of an event in space and time. Betty expected Arthur to reply, using the train as a frame of reference 'three cars down in that direction', but Arthur deliberately chose to adopt the external world as his reference frame. Both answers are equally valid. And, given the speed of the train, it is easy to convert from one frame to another.

The Galilean Transformation

Let us suppose that Arthur started off on his walk when the train was passing Rugby and that he and Betty agree to synchronise their coordinate systems so that, in both frames, this event happens at spacetime coordinates (0, 0). Now Arthur found the refreshment bar three cars, that is 60 m towards the front of the train and bought his coffee 10 minutes after leaving Betty. In this time the train had travelled 30 km (at 50 ms^{-1} or 3 km per minute). With reference to the ground (the 'stationary' frame) the coordinates of the event are (30,060 m, 10 min) but in the 'moving' frame, the coordinates are (60 m, 10 min). In order to translate from the 'stationary' frame to the 'moving' frame, we must *subtract* the distance moved by the train. In other words, if an event occurs at position x in the stationary frame, its positional coordinate in a frame which is moving at a speed v in the positive x direction will be $x - vt$. Of course, we do not expect Arthur and Betty to disagree about the *time* the event took place and in general, the y and z coordinates of an event which occurs in three dimensional space will not change either so the complete set of equations which translates the coordinates of an event from one frame of reference into another which is moving in the positive x direction with respect the the first is:

$$x' = x - vt$$
$$y' = y$$
$$z' = z$$
$$t' = t$$

This set of equation is known as the Galilean transformation, not because he was the first to write them down or even conceive of them, but because he was the first person to realise that it was essentially only *relative* motion that mattered and that, inside a moving frame of reference such as a train or, in his case, a ship, objects would behave in exactly the same way as they do in a stationary frame. It was this realisation that enabled him to conceive of a world which was spinning round once every 24 hours and flying round the Sun once every year without just leaving everything behind like leaves off a lorry.

The principle of Galilean Relativity can be stated loosely as follows: *it is impossible to determine the speed of a train (or planet or space ship) by performing mechanical experiments contained wholly within the train (planet or space ship).*

It is important to include the word 'mechanical' in that statement because there is, potentially, one way in which you might be able to detect absolute motion through space and that is by measuring the speed of light. If you found out that light travelled faster in one direction than in the opposite direction, you would naturally conclude that you were moving through space in the direction in which the speed of light was smallest (because in this direction, your speed would be subtracted from the speed of light).

Now when he was a young man, Einstein has wondered what it would be like to travel in a space ship beside a ray of light. Could you actually make light apparently travel backwards by travelling faster than light? Later, Einstein realised that the theory of electromagnetic waves worked out by the brilliant mathematician and physicist James Clerk Maxwell expressly forbade light from travelling at any other speed than 300,000 km per second and that led him to speculate what would happen if it was impossible to determine your state of motion by *any* experiment at all. What if the speed of light was constant in *all* frames of reference? What then? These speculations led to a bizarre world in which rulers shrink and clocks go slow; fathers can return from a journey to find themselves younger than their sons; long poles can fit inside short barns and pennies can bend as they slip trough holes too small for them to fit.

Normally, in everyday life, these effects are too small to notice because it is necessary for the relative motion between the relevant frames of reference to be a sizeable fraction of the speed of light. But, owing to the extreme accuracy with which GPS satellites have to regulate their clocks, the effects of both the Special Theory of Relativity and the General Theory (which includes the effects of gravity) have to be taken into account in order to enable your smartphone to know where it is so it is well worth while trying to understand how the apparent paradoxes mentioned above arise and how they are resolved.

The Lorentz Transformation

When we were describing the Galilean Transformation in the previous chapter, we made the important assumption that, although Betty and Arthur choose to differ about *where* Arthur bought his coffee, they both agree about *when* he bought it.

But when it comes to trains moving at a sizeable fraction of the speed of light (!), time and distance become inextricably entwined and it turns out that Betty and Arthur have to disagree about both where *and* when the event took place.

The set of equations which relates the coordinates of an event in two coordinate systems in relative motion are known as the Lorentz Transform and is shown in the box below.

$$x' = \frac{x - vt}{k}$$
$$y' = y$$
$$z' = z$$ [3]
$$t' = \frac{t - v/c^2 x}{k}$$
$$\text{where } k = \sqrt{1 - v^2/c^2}$$

where c is the velocity of light.

It is assumed that the motion lies along the X axis so the y and z coordinates are unaffected. It is easy to see that if v is much smaller than c, $k = 1$ and the equations reduce to the familiar Galilean transformation.

A simple scenario

Although, in order to appreciate the true beauty and symmetry of these equations, we must use some elementary algebra, I am always happier to start with a few simple numbers so, following a long tradition which started with Einstein himself.[4] I ask you to imagine a train 100m

---

3   In most text books, the symbol λ is used instead of k where $λ = 1/k$.. λ extends from 1 to ∞ whereas k lies between 1 and 0.

4   Einstein: *Über die spezielle und allgemeine Relativitätstheorie*, Verlag von

long rushing from West to East through a station with a platform 100 m long at a speed of $v = 60$ m/s (about 135 mph). At the back of of the train is a guard and at the western end of the platform stands the station master. We shall regard the instant at which the guard passes the station master as the origin of all our measurements in both space and time. At the other end of the platform (i.e. at a distance $x = 100$m from the station master) a porter is carrying some cans of paint and exactly 1s later he happens to spill one of the cans all over the platform and the passing train. The question is, where on the train will the paint splash appear?

The calculation is easy. In the time $t = 1$s between the two observers starting their watches and the accident with the paint, the train has travelled $1 \times 60 = 60$m and so the paint splash will be found $100 - 60 = 40$m along the train (remember, we are taking the guard as the origin of our moving frame of reference).

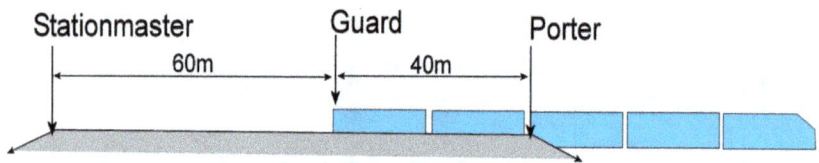

A non-relativistic train passing through a station

Using letters instead of numbers we find:

Distance moved by train $d = vt$

Position of paint mark $x' = x - vt$

You will recognise this as the main equation of the Galilean transform described earlier.

A more interesting scenario

Suppose that, instead of a paint spillage, the station master fires a pistol at the instant the guard passes him. At the other end of the platform there is a microphone which, when it is triggered by the sound of the gun, fires a paintball at the train thereby making a permanent mark on the carriage. The question is – how far along the train will the paint mark show now?

Friedrich Viewhweg & Sohn, 1916

In order to make the calculations as easy as possible we shall assume that sound travels at 100 m/s. (Perhaps it was a very cold day!) Obviously at this speed the sound will take exactly 1s to travel down the platform and the calculation is exactly the same as before. It is, however, worth considering the situation from the point of view of the guard. He sees the platform whizzing by backwards and (if sound waves were visible) he would see the sound waves having to struggle against the wind making headway at a mere speed of $100 - 60 = 40$ m/s relative to the train. On the other hand, the destination (i.e. the microphone trigger) is travelling towards the struggling sound wave at 60m/s so the combined velocity of approach is still 100 m/s and the sound still takes exactly 1s to reach the microphone – but it has only advanced 40 m up the train in this time.

A Relativistic scenario

Now suppose that, instead of a sound wave, we use a light wave. Of course, light travels a million times faster than sound so, to keep the numbers simple and comprehensible, let us also pretend that light travels at a mere 100m/s. Instead of a microphone we have a photoelectric cell but as before this triggers a paint ball marker. At first sight, nothing seems to have changed. The light still takes 1s to travel down the platform so the paint mark ought to be 40 m along the train as before. But when the train is examined in the sheds later, it is found that the paint mark is actually 50 m up the train. How is this? What has gone wrong?

The answer is that light waves do not behave like sound waves. Sound waves travel through a medium – air; and the velocity of sound always has to be measured with respect to the medium through which it travels. That is why, to the guard on the train, the sound waves appear to be travelling more slowly than they should because of the headwind. Light waves, however, travel through space, not air. And since it is impossible to measure a velocity with respect to empty space, it turns out that light *always travels at the same speed* with respect to any observer, however they may be moving with respect to other bodies or observers. This is the fundamental principle on which the theory of Special Relativity is based – the laws of physics, including the laws which govern the speed of light, are exactly the same for all (inertial[5])

---

5   The word *inertial* here indicates that observers in accelerated rockets or

observers.

Now at first sight it would seem that this assumption must lead to contradiction and inconsistency; but this is not the case. It does, however force us to abandon the two assumptions we made earlier namely that measurements of time and distance are unaffected by speed. Let us see why.

In our calculation above, we assumed that the train remained the same length as it passed through the station. We must abandon this assumption. Let us suppose instead that the train is lengthened or shortened by a factor k. (We shall find that with respect to the station master (but not the guard) the train is in fact shortened and that k is, in this situation equal to 0.8). The station master is correct in saying that the light takes 1 s to travel the length of the platform and that in this time the train moves 60 m. He is also correct in deducing that at this instant the back of the train is 40 m from the Eastern end of the platform; but if the train is shortened by a factor k (where $k < 1$ remember), that 40 m will, in fact, contain *more* than 40m of train. It follows that we must *divide* by k to get the correct answer.

In symbols the calculation goes like this:

Time taken for light to travel x m
$$t = x/c$$
Distance moved by train in this time
$$= vx/c$$
Distance of paint mark along the train
$$D = \frac{x - vx/c}{k}$$
$$x' = \frac{x(1 - v/c)}{k}$$

But how do we know what k equals?

To work out k we must look at the situation from the point of view of the guard on the train. To him, the train is stationary and it is the

---

rotating space stations or gravitational fields are excluded. To include such observers we need the General Theory of Relativity.

platform which is travelling backwards at a speed of 60 m/s. In addition, from his point of view, it is not the train which is lengthened or shortened, it is the *platform* which now has length $k \times 100$ m (remember $k < 1$). Now, the light ray is approaching the photocell at a combined speed of $100 + 60 = 160$ m/s. To the guard on the train the ray therefore reaches the photocell in $100k/160$ s and travels a distance of $100 \times 100k/160 = 62.5k$ m. This is the point of the train where the paint will be found.

Now obviously, whatever may happen to clocks and rulers while objects are in motion, the two observers must eventually agree where the paint ball actually hits the train so we have $62.5k = 40/k$ from which we obtain $k = 0.8$.

Repeating this calculation using letters instead of numbers we find that:

$$c \times \frac{xk}{c + v} = \frac{x(1 - v/c)}{k}$$
$$k^2 = (1 - v/c)(1 + v/c)$$
$$k^2 = 1 - v^2/c^2$$

so

$$\boxed{k = \sqrt{(1 - v^2/c^2)}}\ \text{[6]}$$

If you found this argument a bit hard going, think of it this way: the station master thinks that the paint mark is further up the train than expected because the train is shortened; the guard on the train reaches the same conclusion because, although he sees the light ray and the detector approaching each other very rapidly (at 160 m/s) the platform is shortened by just the right amount to cause the paint to be spilled 50m[7] along the train.

The Lorentz Transformation

Let us go back to the first scenario with the accidental spillage. You will recall that (according to the station master) the porter spills the paint 1s after the train enters the station and the coordinates of this event relative to the platform are (100, 1). We need to work out what the

---

6   Putting $v = 60$ and $c = 100$ we get $k = \sqrt{(1 - 0.6^2)} = 0.8$
7   $62.5 / 0.8 = 40 \times 0.8 = 50$

coordinates of this event are relative to the train.

We already know the answer in the case of the spatial coordinate. In a time t the train travels a distance vt. The remaining distance along the platform is $x - vt$ and the corresponding distance along the train is this distance divided by k so:

$$x' = \frac{x - vt}{k}$$

To work out the temporal coordinate we need to look at the situation from the point of view of the guard in the train. To him, it is the platform which is shortened, not the train so from his point of view, at the instant he passes the station master, the porter is only $100k = 80$m down the platform. He sees the porter travelling backwards at 60 m/s, spilling the paint on the train when he reaches the 50m point along the train. The time between these two events is $(80 - 50)/60 = 0.5$s. This is the temporal coordinate we seek.

The equivalent algebra is as follows: the x coordinate of the porter when the train enters the station

$$= k \times x$$

when the paint is spilled (eqn 4)

$$= \frac{x - vt}{k}$$

so the temporal coordinate is

$$t' = \frac{kx - (x - vt)/k}{v}$$

At first glance this expression does not look very pretty but watch what happens when we sort it out a bit. Multiplying top and bottom by k

$$t' = \frac{k^2 x - x + vt}{kv} = \frac{x(k^2 - 1) + vt}{kv}$$

Now since $k = \sqrt{1 - v^2/c^2}$, $k^2 - 1$ equals $-v^2/c^2$. Cancelling out v we get the equation we are looking for:

$$t' = \frac{t - v/c^2 x}{k}$$

Putting $t = 1$ and $x = 100$ we find that $t' = 0.5$.

The constancy of the speed of light

To check that these formulae are consistent we shall use them to see how both the station master and the guard come to the same conclusion about the speed of light.

In the station masters frame, the paint spilling event has coordinates (x, t) so he calculates that the speed of light is $c = x/t$.

In the guards frame, the paint spilling event has coordinates (x', t') so he calculates that is

$$\frac{x'}{t'} = \frac{(x - vt)/k}{(t - v/c^2 x)/k}$$

But the guard is well aware that in the station masters frame $c = x/t$ so he can substitute $x = ct$ into this:

$$\frac{x'}{t'} = \frac{(ct - vt)/k}{(t - v/c^2 ct)/k} = \frac{(c - v)t}{(1 - v/c)t} = c$$

So in spite of the fact that the station master and the guard disagree about where and when the paint spilling event occurred, they both agree that the light beam travelled at the same speed in their different frames of reference.

The constancy of the speed of the train

From the measurements that they have made, both parties can calculate the speed of the train. The station masters calculation goes like this: Since the train is travelling at a speed v, when the paint is spilled, the back end of the train is $100 - v$ metres up the platform. But later inspection shows that the paint was a distance $D = 50$ m up the train. It follows that the factor k is equal to $(100 - v)/50$. This gives the equation

$$\frac{100 - v}{50} = \sqrt{1 - \frac{v^2}{100^2}}$$

which boils down to

$$v^2 - 160v + 6000 = 0$$

This equation has two solutions: $+60$ and $+100$. The first is obviously the correct answer but what can we say about the second? In this case the train is travelling at the speed of light, the train is shortened to zero length and when the paint is spilled on to the train, any value of D is possible.

What about the guard's calculation? He knows that the end of the (shortened) platform and the beam of light are approaching each other with a combined speed of $100 + v$. He also knows that in this time the beam of light travelled 50 m down his train. He can therefore conclude that

$$\frac{100 \times \sqrt{1 - v^2/100^2}}{100 + v} = \frac{50}{100}$$

or
$$v^2 + 40v - 6000 = 0$$

This equation has two solutions: -60 and $+100$. The first answer is negative because the guard is not measuring the speed of the train – he is measuring the speed of the platform. The second is the degenerate result as before.

Later that day the guard and the station master were pondering over the position of the paint on the train. Not knowing anything about Relativity their conversation might have gone something like this:

S'master: I don't understand it! I *know* that the paint was spilled exactly 1 s after you passed me because the platform is 100 m long and light travels at exactly 100 ms^{-1}. I also *know* that the train was travelling at exactly 60 ms^{-1}. So the train *must* have travelled 60 m in that second and that the back end of the train was therefore 60 m up the platform when the paint was spilled. So why is the paint 50 m up the train not 40 m? The *only* possibility is that the train must have shrunk so that 50 m of the train occupied the space of 40 m of platform.

Guard: I don't understand it either but I disagree totally with your analysis. You see, I *know* that the train didn't shrink – after all I was on the train not you! I agree with you that the train was travelling at 60 ms^{-1} because earlier I timed how long it took for the train to pass under a bridge. The time was 1.67 s as you would expect. The first thing I disagree with is your statement that the light took 1 s to travel down the platform. I think it took a lot less than that because the end of the platform and the light beam were approaching each other at a combined speed of 160 ms^{-1}. In fact I can tell; you exactly how long it took because the paint was found 50 m up the train. Now since the speed of light is 100 ms^{-1}, it is obvious that it only took 0.5 s to reach that point on the train.

S'master: But that is ridiculous! My watch clearly showed that it took 1 s to travel down the platform.

Guard: Well your watch must be running slowly- that's all I can say.

S'master: Wait a minute. If you say that it only took the light beam half a second to travel down the platform, the train will only have moved 30 m in that time and the paint ought to be 70 m up the train – not 50! How do you explain that? Eh? Eh?

Guard: Well …

S'master: Got you now, haven't I?

Guard: Hang on – just let me think a minute. The light beam is travelling at 100 ms^{-1} towards the end of the platform but the end of the platform is moving towards the light beam at 60 ms^{-1} so the combined speed of approach is 160 ms^{-1}. Now the platform is 100 m long so the time taken will be 100/160 which is...

S'master: 0.625 s! I knew you were talking rubbish! Your numbers just don't tally! Admit it! You're wrong. There is nothing wrong with my watch – it is the train which shrinks and that's an end to it!

Guard: Okay Okay. Stop crowing. There is something wrong here but I can't quite put my finger on it. You think that my train has shrunk so by rights I should think that your platform has shrunk...

S'master: Oh! don't be absurd! They can't both shrink

Guard: Hang on, I'm nearly there. I *know* that the combined speed of approach is 160 ms^{-1} and I *know* that the time taken was 0.5 s so the platform *must* be $160 \times 0.5 = 80$ m long. That's it! The platform shrinks in exactly the same ratio as the train! The situation is exactly symmetrical! You think my train has shrunk but I think that your platform has shrunk!

S'master: Well – I suppose there is some logic to that. But if we are to be entirely consistent, if, as you say, my watch is running slowly, then by all rights I should think that your watch is running slowly. But we know that when my watch reads 1 s yours reads 0.5 s.

Guard: Yes, you do have a point there. Give me another minute...

S'master: Come on. Time's up. You have had at least five.

Guard: Okay, listen. You said that when your watch reads 1 s my watch reads 0.5s.

S'master: Sure thing.

Guard: But how do we know that?

S'master: Well *I know* that it took 1s by my watch for the light to travel 100 m up the platform and *you say you know* that it took 0.5 s for the light to travel 50 m up the train.

Guard: Yes I know that – but that isn't the same thing as saying that your watch is running twice as slowly as mine because at the instant the paint gets spilled *our watches are not in the same place*!

S'master: What of it?

Guard: Well, you can't put them side by side and compare them directly can you?

S'master: I don't see what that has got to do with anything. After all, we could just get our watches out and compare them here and now and settle the matter.

Guard: I don't think that would settle the matter at all. All this business about shrinking platforms and watches running slow depends crucially on the speed of the train being constant. Since I passed through your station, my watch has turned round and come back the other way so any information about how fast it was going then will probably have been nullified on the return journey.

S'master: Well it all seems crazy to me.

Guard: I agree. But even if we don't understand all of the in's and out's, the proof that something crazy really does go on it here right in front of you – a splash of paint in the wrong place!

S'master: Okay. So lets suppose that you are right and that, to me, the train has shrunk, but to you, the platform has shrunk. What about our watches? If, as you say, to you my watch is running slowly, then surely, to me, your watch is running slowly. Where is the evidence for that?

Guard: Yes, that's a good point. I said earlier that your watch was running at half the speed of mine but, on reflection, I don't think that is quite right. According to your watch, the two events (the triggering of the light pulse and the spilling of the paint) were 1 s apart. Now from my point of view, I saw a light beam travel 80m down a shortened platform in this time at a speed of 100 ms^{-1}. In other words, in my frame of reference the paint was spilled 0.8 s later and your watch is running 80% slower, not 50%.

S'master: How do you reconcile that with the fact that, according to you, the paint was spilled 0.5 s later?

Guard: I am not expressing myself clearly. I shouldn't have said 'in my frame of reference the paint was spilled 0.8 s later' because you are absolutely right – in my frame of reference the paint was spilled 0.5 s later. What I should have said was ' in my frame of reference I calculate that in your frame of reference the paint was spilled 0.8 s later and that since your watch actually read 1 s, I conclude that your watch is running 80% slow'.

S'master: That sounds a bit complicated but I think I see what you mean. What about my views on your watch? Surely, whatever we calculate, to me, your watch must be running faster than mine; so where is the symmetry in that?

Guard: Well, from your point of view, you see a light beam travelling 40 m up a shortened train. You calculate that, from my point of view, this should only take 0.4 s. But I am telling you that my watch actually reads 0.5 s so you conclude that my watch is going 80% slow too!

S'master: Basically what you are saying is that it doesn't make any sense to talk about when, or indeed, exactly where the paint spilling took place. We just have to agree to differ.

Guard: Yes – exactly.

S'master: So is all this business about trains shrinking and watches running slow just an illusion? It doesn't really happen does it?

Guard: Well, that's an good question. You can argue that all this business is just a different perspective on the same events. When you look at a circular table for an oblique angle it looks like an oval – but nobody would claim that the table has actually shrunk in one dimension just because you were looking at it from a different angle. In this case there is an obvious frame of reference (namely the three dimensional space in which the table sits) which is to be preferred over the arbitrary two-dimensional view which we get from a specific location.

But in the case of trains and platforms, there is absolutely no reason why my frame of reference is better than yours or yours, mine. From my point of view, your platform really is 80 m long and all its atoms are really squashed. Everything in your world really does seem to be running slow; your watch is running slow; you breathe more slowly, you will live longer than me etc. etc. These are real effects; not illusions.

Likewise, to you, my train really is shorter than it was when it started its journey and it is my watch which is running slowly. What is more, I can prove to you that the effect is real, not an illusion because – look – our watches do not agree. Mine is a few seconds slow compared to yours.

S'master: But doesn't that just undermine everything you have just said about there being nothing to choose from between our two frames of reference?

Guard: No. Because, remember, it was me who travelled to the end of the line and came back again. Your frame is special because you never changed your speed at any time. Your frame is called an inertial frame. Mine wasn't. That is the difference.

S'master: So there really is no paradox here?

Guard: No, none at all. It all checks out beautifully.

S'master: And Einstein figured all this out while he was twiddling his thumbs in an obscure Patent Office?

Guard: That's right. He did.

When the Guard pointed out that their two watches would disagree when they were reunited he was opening a whole new can of worms, commonly known as 'the Twins Paradox'.

The Twins Paradox

Albert decides to travel in a fast spaceship at 80% of the speed of light to our nearest star Alpha Centauri which is 4 light years away leaving his twin brother Ludwig behind on Earth. Not knowing much about relativity he and his brother expect the journey to take 5 years there and 5 years back. (Time = distance / speed = 4 / 0.8 = 5 years). To his astonishment, Albert finds that the journey takes less time than he thought. 3 years, to be precise. Not finding anything of interest at Alpha Centauri, he heads for home. Arriving back at Earth only 6 years older (a fact verified by the observation that he had only eaten a little more than half the food he had taken with him) an even greater surprise awaits him. His brother Ludwig swears that he has been away 10 years after all - a fact amply proved by Ludwig's new wife and a family of 7 children!

The story is surprising enough but the fact that time proceeds more slowly for Albert than for his brother is not paradoxical. It is, as we have already seen, a straightforward consequence of Einstein's theory of Special Relativity; a consequence which has been verified a thousand times over by a great number of very different experiments to a remarkable degree of accuracy. However much you might find the result unpalatable, it remains a fact and contains no paradox.

So where *is* the real paradox? Let us continue the story. During Albert's absence, Ludwig had found a book about Relativity in the local library and had studied it closely. He had found it so interesting, in fact, that he had faxed a copy to his brother in the spaceship. (By a curious coincidence, he had faxed it at the precise moment when he had reckoned his brother would be arriving at Alpha Centauri - ie 5 years after his departure). In the book he had learned about time dilation and so he knew that Albert would only be 6 years older when he returned. When they were reunited, he was delighted to see his brother in such good health

Albert, on the other hand, was utterly dismayed when he saw his brother. Owing to the time it takes light (and radio waves) to travel, he had received his copy of the book on his way home and had not had time to read it properly. He had, however, got to the bit about time dilation and he had reasoned as follows: 'All motion is relative, according to Einstein, so I might as well assume that I am the stationary one and that it is my brother Ludwig who is travelling at high speed away from me. Since he is in motion (relative to me) his clocks must be going slow (relative to mine) so when I get home I will find that *he* is younger than *me*.' As events were to prove, he was mistaken - but what is wrong with his reasoning?

I have read several explanations of this paradox and I have never been satisfied with any of them. Some juggle with Lorentz transformations and prove it mathematically[8]. Others point to the fact that the situation is not, in fact, symmetrical because Albert undergoes accelerations and decelerations during which the principles of *Special* Relativity do not apply[9] [10]. The first approach does not help if you are not a mathematician and the second is at best misleading and quite possibly wrong. Certainly there is no need to invoke any new principles from Einstein's General theory to explain the paradox and while it is true that the situation is not symmetrical, the paradox arises precisely because there is an unexpected and pleasing symmetry in Einstein's solution which is absent from the non-relativistic expectations of the two brothers.

To explain what I mean, it is necessary to examine very carefully indeed exactly what the two brothers *expect* to happen during the journey assuming that there are no relativistic effects at all. To emphasise that we are discussing a classical, Newtonian, world we shall call our explorer Isaac and our stay-at-home brother, Christian.

Let us suppose that Isaac sets out on Christmas Day on his 5 year voyage to Alpha Centauri. As a parting gift, Christian gives to Isaac a lovely clock which shows not only the time but the date and the year in big luminous letters. 'With this clock,' said Christian, 'you will be able to tell exactly what time it is back home whenever you want.' 'Thank you so much! I have a present for you too.' his brother replied, pulling out a small but beautifully made telescope. 'You can watch me fly away with this and if I put my new clock in the window of the spaceship, you will

8 Feynman, R. *Lectures on Physics*, Addison-Wesley **1963**

9 Davies, P. *About Time.* Viking **1995** 0-670-84761-5

10 Gardner, Martin. *Relativity Simply Explained.* Dover **1996** p. 116

be able to tell what time it is aboard ship whenever you want!'. Both brothers were so delighted with their presents they decided to buy another clock and another telescope so that whenever they wanted to, they could look down their telescopes and see what their brother was doing and what time it was. They also agreed that throughout the voyage they would keep in contact by sending regular Christmas radio messages to each other.

Now both brothers were well aware that light (and radio waves) travel at a finite speed. They were quite prepared, therefore, to accept that, when looking through their telescopes, the brother's clock would not look as if it was saying the same time as their own; they also knew that they would receive their brother's messages long after they were sent. In order to keep track of things, each brother prepared a graph showing when (as shown on *his* clock) he would expect to receive his brothers messages. This is what the Earth-bound Christian's graph looks like.

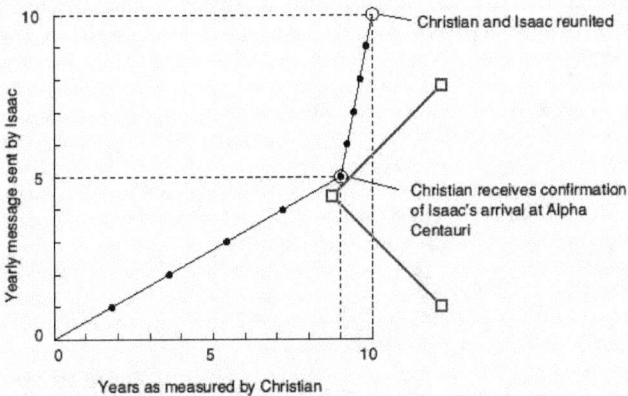

On the X axis we have Christian's time scale in years from 0 to 10. Since it will take 5 years for Isaac to reach Alpha Centauri and 4 years for the message to return home, Isaac's *fifth* Christmas message in which he confirms his arrival at his destination, will arrive *nine* years after his departure. It follows that the first 5 messages will be received at 1.8, 3.6, 5.4, 7.2 and 9.0 years. This is shown on the graph by plotting a series of dots against the relevant year. The remaining messages come a lot closer together because Isaac is on his way home. In effect, the graph plots Isaac's time (as seen by Christian through a telescope) against

Christian's time. (Remember - we are assuming that there are *no* relativistic effects, only effects due to the finite speed of light.)

Now what does Isaac's graph look like? The situation is a little more complicated here because Isaac can make one of two assumptions. The first is to accept that it is he who is moving through space and that when he is travelling away from Earth, messages take longer to reach him because the radio waves have to catch him up as he speeds away from the transmitter. (Once again, I must remind you, we are not talking *Relativity* here!), On the outward journey, Isaac is receding from Earth at 80% of the speed of light. One year out (when Christian sends his first message) he is 0.8 light years away. The radio waves are chasing him at the speed of light but the closing speed is only 0.2 light years per year. The radio waves will therefore take 0.8 / 0.2 = 4 years to catch him up and will therefore reach him at the precise instant that he reaches his destination. On the return leg, the closing speed between Isaac and the incoming messages is 1.8 times the speed of light. At this closing speed the time taken for a radio wave to meet the homecoming Isaac will be 1/1.8 = 0.55 years so Isaac will receive the remaining 9 messages in 9 * 0.55 = 5 years.

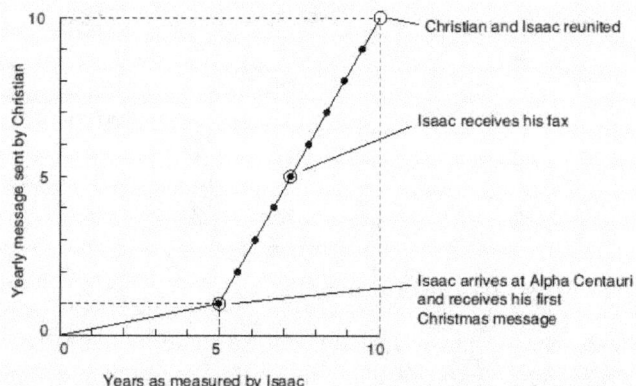

It will be noted that, although the two graphs are rather different, both brothers expect to be exactly 10 years older when they meet again! It is also worth noting that on the outward journey, both see their brothers clocks apparently running slow - but by different amounts. Christian receives 5 messages in 9 years, a time factor of 5/9 or 0.555. Isaac receives only one message in 5 years, a time factor of 0.2. (This difference is due to the fact that Christian experiences the Moving

Source Doppler effect while Isaac sees the Moving Observer Doppler effect.[11]) On the return leg, the time factors are 5 and 1.8 respectively.

But suppose that Isaac were to insist that it was *he* who was stationary and that it was Christian that was moving. Then, of course we would be back to the Moving Source Effect and Isaac would expect to see the same effects that we previously worked out for Christian. In this scenario you must imagine Isaac in his stationary rocket. For 5 years, Earth recedes and Alpha Centauri approaches, then for 5 years Earth approaches and Alpha Centauri recedes. The following graph shows when Isaac would receive his brother's messages.

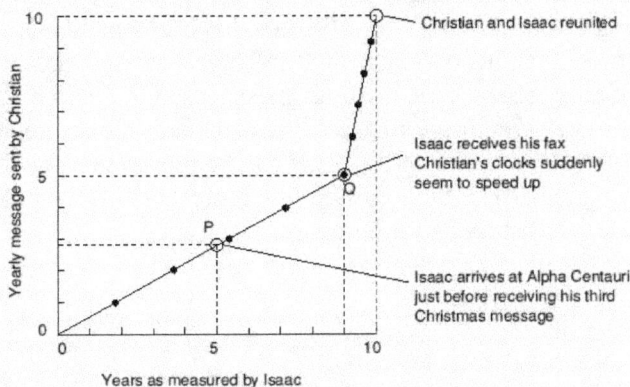

Note that under these circumstances, events turn out differently. Instead of receiving his first Christmas message when he arrives at his destination, Isaac will receive his third a short time later; and the changeover point when his brother's clocks change from going slow to going fast occurs not when he arrives at Alpha Centauri, but several years later, when he receives his fax.

It is, I hope, clear from this that in a non-relativistic world, the timing of events depends on who is stationary and who is moving (relative to the supposed medium in which light travels - the aether). The miracle of the Special Theory of Relativity is that it doesn't matter who is moving and who is stationary, the results always turn out to be the same - only the results aren't always what you expect!

So how does Special Relativity achieve the miracle of allowing both Albert and Ludwig (the relativistic pair) to see their brother's clocks

11 An explanation and derivation of these will be found in the Appendix.

going slow, and yet allow one to be older than the other when they meet? The answer lies in contraction of length and the dilation (ie expansion) of time.

As we have seen in the previous chapters, the special theory of relativity predicts that, as measured by any observer who considers himself stationary, moving clocks will run slow and that moving metre rulers will appear shorter by a factor k which is the same in both cases and is equal to $\sqrt{1 - v^2/c^2}$. In the case we are considering the factor is equal to $\sqrt{1 - 0.8^2} = 0.6$. When Ludwig learned about time dilation, he realised that his astronaut brother Albert would only be 6 years older when he got home because for 10 years, his brothers clocks would be going slower than his own. During the 10 year voyage, Ludwig therefore only receives 6 Christmas messages and, *compared to his expectations in the absence of Relativity*, Albert's clocks appear to be running slow by a factor of 0.6. This can be represented by the following graph. (The faint dashed line represents Ludwig's (non-relativistic) expectations for comparison)

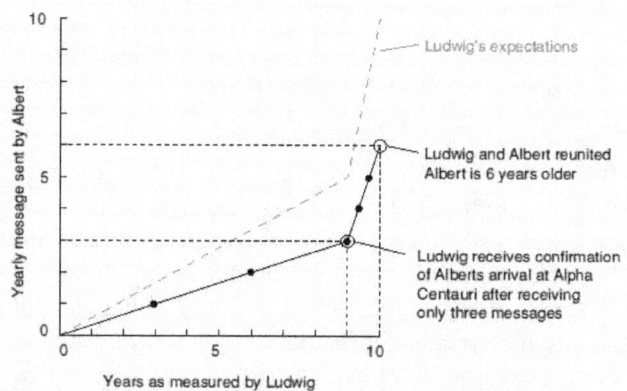

The total time factor as measured by Ludwig on the outward journey is 3/9 or 0.33 and is the product of the Doppler shift factor 5/9 and the time dilation factor 0.6.

Now why does Albert only send three messages before his arrival? The answer is that, to Albert, he is, by definition, stationary (*all observers are by definition stationary in their own frame of reference!*) and it is the Earth which really is receding and Alpha Centauri which really is approaching! To Albert, therefore, the Earth and Alpha Centauri constitute a huge cosmic ruler which was 4 light years long before he started moving. Here comes the crucial point. Because he is travelling so fast, *the length of this ruler shrinks* to 0.6 of 4 which is *2.4 light years* so it only takes 2.4 / 0.8 years to go by at a speed of 0.8 light years per year. This is, of course, exactly 3 years. The same is true on the way home. *That* is why Albert makes the round trip in 6 years. He doesn't have to go as far as he thought!

What about the messages that he receives from his brother? Because of the exact symmetry of the relativistic situation, *Albert must see exactly the same total Doppler shift in Ludwig's messages as Ludwig sees in Albert's*. On the outward journey, Ludwig got 3 messages in 9 years. Albert must therefore receive exactly 1 message in the 3 years it takes him to reach Alpha Centauri. By the same token, he will receive 9 messages on the return journey. His graph therefore looks like this (as before, the faint dashed line represents Albert's (non-relativistic) expectations based on him being stationary and Ludwig receding from him):

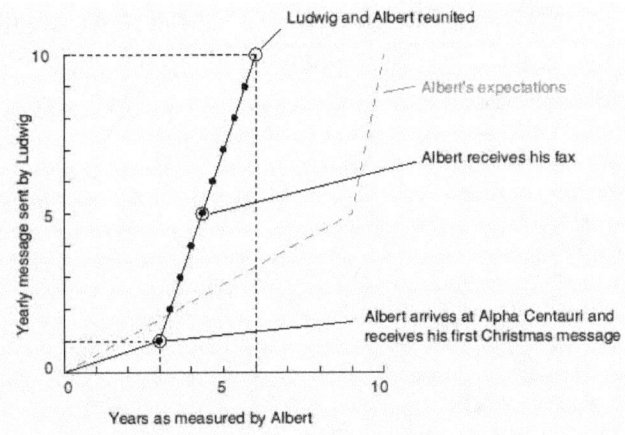

We are now at long last, in a position to see clearly in what sense it is true for *both* brothers to claim that the other brother's clocks are running slow and yet for the two brothers to have different ages at the end of the voyage. For Ludwig, the effects of time dilation on his brother are immediately obvious. Albert is 4 years younger than he 'ought' to be! For Albert, the argument is a little more subtle. Once again, because of the exact symmetry of the relativistic Doppler shift effect, Albert *expects* to see the same pattern of messages as his brother - ie 5 messages in the first 9 years and 5 in the last year. What he *actually* sees is 1 in the first 3 (a slower rate) and 9 in the next 3 (also a slower rate). In short, the *gradients* of both Ludwig's and Albert's graphs are identical. Both see the other as being both Doppler shifted and time dilated. What is different is that for Ludwig the slow rate lasts for 9 years and the quick one for 1 year resulting in Albert being only 6 years older when he gets home. For Albert, the slow rate only lasts for 3 years and the quick rate for all of another 3 years resulting in Ludwig being 10 years older when they are reunited.

As they say in the ad - 'it's all worked out beautifully.'

Perhaps the most important thing to take home from all of this is that the resolution of the Twins Paradox has nothing at all to do with accelerations and decelerations, nor has it anything top do with General Relativity or the warping of spacetime; It is simply a result of the combined effects of the classical Doppler shift, length contraction and time dilation. The asymmetry between the two twins arises simply because it is Albert who swaps from one inertial frame of reference to a different one – while his stay-at-home brother remains in the same inertial frame throughout.

The Twins Paradox revisited

'All these graphs – its just too confusing!' exclaimed Arthur having just finished reading the previous chapter. 'What with Doppler shifts, time dilation and length contraction – isn't there a simpler explanation?'

'Well, it depends what you mean by simpler.' said his elder sister, Betty. 'Some people like real figures and pictures, others are happier with a more abstract approach.'

'Well, I am not sure I like the sound of that any better but you might as well give it a try.'

'Okay – but you are going to have to let me start somewhere.'

'What do you mean?'

'Well – the author just assumed the formulae for length contraction and time dilation. If we are going to start somewhere else, we must make some similar assumptions.'

'Fair enough.' agreed Arthur.

'First lets start with the concept of spacetime. Any *event* (such as your birth or the explosion of a distant supernova) takes place at a certain location and at a certain point in time. For example, you could specify your birth in spacetime by stating the planet you were born on, the latitude and longitude of your birthplace and the time and date. In general you need three spacial coordinates and one temporal one to fix the location of any event in spacetime. In fact we can refer to a set of 4 coordinates (t, x, y, z) as an *event* whether or not anything actually happens there.'

'Yes, I get that.'

'Good. But we shall be interested in the relationship between two such sets of coordinates – Albert's and Ludvig's – which are in relative motion. We shall assume that at the instant Albert sets off on his travels, these two coordinate systems coincide. i.e. the event of Albert's departure is (0, 0, 0, 0) in both systems and that Albert moves off in the X direction. What this means is that the Y and Z coordinates of Albert's journey are always zero so we shall forget them from now on.'

'So in Ludvig's system, Albert travels 4 light years in 5 years and then comes back again while in Albert's system, he remains stationary the whole time. Is that right?'

'Yes, on the outward journey; but when Albert turns round he has to abandon his outward coordinate system and step on a different system for the journey home. So there are really *three* coordinate systems to consider. Ludvig can use one system for the whole time but Albert cannot. *This* is why there is an asymmetry in the situation.'

'Yes I see.' said Arthur. 'But can you show that it is Ludvig and not Albert who is older when the brothers are reunited?

'Yes, but to do this we need to employ some principle which embodies the basic assumption on which Special Relativity rests – namely the constancy of the speed of light.'

'And what might that be?'

'It takes the form of a set of equations which relates the coordinates of an event in one system (x, y, z, t) with the coordinates in another (t', x', y', z') which is in relative motion (along the X axis) with speed v. These equations are:

$$x' = \frac{(x - vt)}{k}$$
$$y' = y$$
$$z' = z$$
$$t' = \frac{(t - v/c^2 x)}{k}$$
$$\text{where } k = \sqrt{(1 - v^2/c^2)}$$

'Ah! I can see the length contraction and time dilation factor in there!'

'Absolutely.'

'But how do we use these equations? What do they do?'

'Well, if you know the coordinates of an event in one frame, you can instantly work out its coordinates in another. For example, in Ludvig's frame, Albert's departure is $(0, 0, 0, 0)$ and you can easily verify that when $x = 0$, $y = 0$, $z = 0$ and $t = 0$, then so do x', y', z' and t'.'

'I see.' said Arthur.

'See if you can work out the coordinates of Arthur's arrival in Ludvig's reference frame.'

'Well, according to Ludvig, his brother reaches Alpha Centauri (a distance of 4 light years) 5 years later (travelling at 80% of the speed of light). So $x = 4$, y and z are both zero of course, and $t = 5$. i.e. Albert arrives at his destination at (4, 0, 0, 5) in Ludvig's frame. Am, I right?'

'No. For technical reasons it is usual to put the time coordinate first so the correct answer is (5, 4, 0, 0) '

'OK.'

'Now work out the coordinates in Albert's frame.

'All right. First $k = \sqrt{1 - 0.8^2} = 0.6$ doesn't it? So $x' = (4 - 0.8 \times 5)/0.6$ which equals... Oh! Of course. It equals zero doesn't it? Because Albert's X coordinate is always zero in his frame!'

'Correct. But what about his t coordinate?'

'OK. $t' = (5 - 0.8 \times 4/c^2)/0.6$ But what do I put in for c?'

'Well, since you have been using light-years for the distances and years for the time, c is just 1 light-year per year so put in 1.'

'I see. Well that comes to $t' = (5 - 0.8 \times 4)/0.6$ which is -3 years exactly!'

'Precisely. Albert is 3 years older when he arrives at Alpha Centauri – not 5 years.'

'So if I put $x = 0$ and $t = 10$ into the formula I should be able to show that he is only 6 years old when he gets back home shouldn't I?'

'Hang on a minute..' said Betty.

'No don't spoil my fun – let me work it out: $t' = (10 - 0.8 \times 0)/0.6$ which equals – wait a mo! What's gone wrong? That comes to 16.7 years not 6!'

'I tried to tell you.'

'What?'

'You have forgotten what I said earlier. When Albert turns round he has to abandon his outward coordinate system and hop onto another one travelling in the opposite direction. So you can't just plug the new numbers into the old equations.'

'So what equations can we use?' asked Arthur.

'Well imagine a kind of mirror image of Albert reflected in a magic mirror on Alpha Centauri. When Albert departs from Earth his image is 8 light years away and as Albert travels towards the star, his image converges on him and they return to Earth together. What we must do is write down the equations which relate Ludvig's system to that of Albert's *image*.'

'How do we do that?'

'The first thing to note,' said Betty, 'is that Albert's image travels in the opposite direction so v is -0.8 not 0.8. Second we must arrange it so that when $t = 0$ and $x = 8$ (Albert's image is 8 light years from Earth at the start) then $x' = 0$. These equations will do the trick:

$$x' = ((x - 8) + 0.8t)/0.6$$
$$t' = (t + 0.8(x - 8))/0.6$$

'Yes, I can see that when $t = 0$ and $x = 8$ then $x' = 0$; and I see that you have changed the sign of v.'

'Now put in Ludvig's coordinates for Albert's return i.e $x = 0$ and $t = 10$.'

'Here we go - $t' = (10 + 0.8 \times (0 - 8))/0.6$ which is, hang on, yes! It is 6 years!'

'Excellent! Are you any more satisfied?'

'Well, to be honest, not really. I am still finding it difficult to accept that time actually goes more slowly for Albert than it does for Ludvig.'

'That is because you still haven't grasped the subtlety of time dilation. Time does *not* go more slowly for Albert. For him, time proceeds at its normal rate. It is the *distance* which he has to travel which is shorter than he expects. For Ludvig, the distance Albert travels is the same – namely 4 light years – but Albert's clocks and metabolic rate and everything else are going slow so he takes less *time*.'

'I suppose so.'

'Put it this way. Both Albert and Ludvig travel through spacetime from (in Ludvig's frame) (0, 0, 0, 0) to (10, 0, 0, 0). But they take different routes to get there. Ludvig passes through (5, 0, 0, 0) while Albert passes through (5, 4, 0, 0). It is a bit as if Ludvig travels from A to B directly but Albert goes via C. In ordinary space, the distance between two points is calculated using Pythagoras' theorem:

Distance between two points $= \sqrt{\Delta x^2 + \Delta y^2 + \Delta z^2}$

(where the Δ sign means 'change in') but in spacetime you have to calculate the distance (or 'interval') between two events differently:

Interval between two events $= \sqrt{t^2 - (\Delta x^2 + \Delta y^2 + \Delta z^2)/c^2}$ [12]
where c is the velocity of light (in this case equal to 1 light year per year)

The interval between (0, 0, 0, 0) to (10, 0, 0, 0) is 10 years.

Now the interval between (0, 0, 0, 0) and (5, 4, 0, 0) is $\sqrt{5^2 - 4^2} = 3$ years. So is the interval between (5, 4, 0, 0) and (10, 0, 0, 0). So the total interval between (0, 0, 0, 0) and (10, 0, 0, 0) going via (5, 4, 0, 0) is only 6 years. It may seem odd that going the 'long way round' takes a shorter time – but that's relativity for you!

---

12 See page 57 for the reasons why I have chosen to define interval in this way.

The Pole in the Barn Paradox

'OK.' said Arthur one day. 'I just about get time dilation and the twins paradox – after all, if you travelled directly from London to Leeds by car, your odometer would read about 200 miles, but if I did the same journey via Birmingham, it would be surprising, to say the least, if my odometer read the same as yours.

'Yes, indeed, except that your odometer would read *more* than mine but in the case of the twins, the one who does the 'extra distance' (through spacetime) ends up being *younger* than the stay-at-home twin.'

'Yes, but that's because 'distances' through spacetime are measured with a minus sign in front of the distance terms isn't it?'

'Absolutely! I am impressed!'

'Thanks. You see, I am not completely stupid.' said Arthur. 'So moving clocks go slow. Yes, that makes some sort of sense. I know that it is not that the motion itself causes the clock to slow down in some sort of physical sense – because, to you, it is *my* clocks which are going slow; its just an illusion isn't it?'

'Well, not really. You can't just write off the effects of Special Relativity as some kind of illusion. It is not the case that your clocks just *look as if* they are going slowly – they *really are* going slowly. And, of course, to you, my clocks *really are* are going slow. But there is no way we can put the clocks side-by-side, as it were to compare them directly because, while we can reset them to zero as they fly past each other, by the time we want to compare them a minute or a day later they are miles apart.'

'I guess so. But that sounds like a bit of a cheat to me. It is like two kids arguing over whether their bit of cake was bigger after they have eaten it!'

'That's a rather nice image! I like it.' said Betty.

'Hey – that has given me an idea.' Arthur continued, 'OK we can't put two clocks side by side but we can put two poles side by side and compare them as they go by. According to you, my pole should be shorter than yours – but your pole should be shorter than mine! How can that be?'

'Well, it's just the same as the clocks.'

'No it isn't just the same.' said Arthur. 'Look. You know that old pole of my uncle's[13] that is hanging outside the barn in farmer Jarvis' field. We can't get it into the barn because the pole is 5m long while the barn is only 4m long. But if I were to run through the barn carrying the pole at, say, 80% of the speed of light (!) according to you the pole would shrink and it might even fit into the barn!'

'You are quite right. Yes it would. At a speed of $0.8c$ the pole will be shrunk by a factor of $\sqrt{1 - 0.8^2} = 0.6$ so the pole will be exactly 3m long and will easily fit inside the barn.'

'But that's ridiculous!' exclaimed Arthur.

'No it isn't. Because you are going so fast, the pole is contracted and now it will fit into the barn – according to me at any rate.'

'Well either the pole fits into the barn or it doesn't. Surely there can be no argument about that.'

'Well...'

'OK put it like this. Suppose that at the instant the front of the pole hits the back wall of the barn, you slam the doors shut. Now either you succeed in shutting the door or you don't. If you are right, the door will shut but if I am right, at least a metre of pole will be sticking out of the barn.'

'No, No...'

'Ha! Got you! The situation is even worse than I thought!' said Art6hur gleefully, 'From my point of view running alongside the pole, it is not the pole which has shrunk – it is the barn! The pole is still 5m long but, to me, the barn is only $4 \times 0.6 = 2.4$m so over half the pole will still be outside the barn!'

'Yes, but you are forgetting that not only does special relativity affect poles and clocks, it also affects what different observers regard as simultaneous.'

'Yes, I know that – but you can't have a situation where one observer sees an intact pole inside a barn while another sees a broken pole, half inside and half outside.'

'True.'

---

13 Arthur's uncle was a contender in the 1956 pole-vaulting championships in Melbourne

'So which is it? Who is right? You or me?'

'I am, of course.'

'Well explain it to me then.'

'The flaw in your argument is to be found in the phrase '*at the instant the front of the pole hits the back of the barn.*' Suppose I rig up an electronic sensor which detects when the pole hits the wall. This sensor sends a signal to the door as fast as possible (i.e. at the speed of light) to cause it to shut. Lets consider the situation now from your point of view:

You are stationary, holding a pole which is 5m long. Racing towards you is a barn which is contracted to 2.4m. At the instant the front of your pole crashes through the back wall of the barn, 2.6m of pole has yet to enter the barn and at the same time a flash of light (the signal from my sensor) starts to travel towards the door – but because the signal cannot travel faster than light, and because the barn is already travelling at 80% of the speed of light, the signal can only overtake the barn at a relative speed of 20 %. This means that in the time it takes the signal to travel the length of the barn (2.4m according to you) the barn itself has travelled 5 times further i.e.12m so, by the time the signal reaches the door, the back of your pole is well inside the barn.'

'Yes, but that doesn't really answer my question because by the time the door slams shut, not only has the back of the pole passed through the door, it has actually passed through the whole barn! At no time is the pole wholly inside the barn, which is what you seem to be maintaining.'

'I agree that, to you, there is no instant when the pole is wholly inside the barn – but to me, there is. Actually, I haven't really done justice to my argument because, with the arrangement I have suggested, the door slams shut long after the instant which we are interested in. Let me suggest an alternative scenario. Suppose I put my sensor on the door itself but *delay the shutting of the door* by exactly that time which I calculate it will take for the front of the pole to travel the length of the barn. (Since the barn is 4m long and the pole is travelling at $0.8c$, this will be $5/c$ s)'

'OK lets calculate where the pole will be after this time according to me. The barn is travelling towards me at $0.8c$ and I see your clock start to tick at the instant the barn reaches the front of the pole. $5/c$ seconds later, the barn has moved $0.8c \times 5/c$ metres which is.. let me see... 4m! That still leaves 1m sticking out of the barn like I said all along!'

'Hang on a minute – you have forgotten one vital factor!'

'What's that?' asked Arthur.

'Time dilation. To you. My clocks are going slow. Remember?'

'Oh, yes.'

'In fact every second recorded by my clock is 1.67 of your seconds so the barn will have moved that much more than 4m – 6.67m in fact; more than the length of your pole.'

;Yes, I had forgotten that. But I am still not satisfied. It is all very well proving that by the time you get round to shutting the door, the back of the pole is inside the barn, that still doesn't prove that *the whole of the pole is inside the barn*. To me the front of the pole is sticking out of the other end!'

'Yes, there is no way round it. To me, the whole of the pole is inside the barn; to you, the pole is sticking out of one end or the other.'

'Hang on.' said Arthur, 'I have thought of something else. Suppose that at the instant the front of the pole hits the end wall the pole stops dead. According to you there will be no problem in shutting the door whichever sensor you use. What is going to happen now? Will the pole suddenly expand to its original length and puncture its way through the ends of the barn? And what would this look like from my point of view?'

'That's a good question. What exactly will happen? The first thing you have got to realise is that nothing, *nothing*, can travel faster than light. When the front end of the pole hits the end wall, there is no way that the back end of the pole can know about this event so it must carry on moving forwards as if nothing has changed. So the pole cannot 'stop dead' as you put it. What happens is that a shock wave travels down the pole at a speed which is, of course, less than the speed of light and by the time this shock wave has reached the back of the pole, the pole is completely inside the barn.'

'So what actually happens to the pole?'

'As the shock wave travels along the pole it physically destroys the pole and at the end of the day what we both see is the shattered remains of the pole wholly inside the barn.'

'Yes, I see. Sort of anyway. But lets go back to my original point – are the effects of special relativity just illusions or are they real? I still maintain that they are just illusions caused by the finite speed of light. I see that, to you, the pole is shorter than the barn – but we both know that the pole is *really* longer than the barn.'

'Yes we both agree that *when the pole hangs on the side of the barn* the pole is the longer but when the pole and the barn are in relative motion, then from my point of view, the pole *really is* shorter than the barn.'

'No you can't really mean that.' complained Arthur. 'Just changing your point of view doesn't change what is really real! I mean, look at this pencil (holding up a pencil pointing towards Betty). To me it looks like a pencil but to you it looks like a tiny disc. But we both know that it really is a pencil even though it looks different from different angles.'

'Yes I do take your point – but there is something about the effects of special relativity which make them qualitatively different from other 'illusions' like the apparent bending of a stick placed in water or so-called 'fictitious forces' which appear to act on an object placed in a rotating frame of reference. Whether you call these effects 'real' or not is basically up to you but the thing which sets the effects of special relativity apart is that while in the other cases there is an obvious frame of reference which we can all agree on as being the ultimate 'reality', in SR all frames of reference are equally valid. Obviously we do not want to say that there is no such thing as reality. But the only alternative is to say that all observers carry around a different reality with them.'

'So you are saying the the shortening of my pole is not an illusion – it is really real?'

'Yes – but you mustn't misunderstand me. All the measurements which I can make with my sophisticated clocks and measuring devices tell me that your pole *really is* 3 m long. But I am perfectly happy to accept that all the measurement *you* make on the pole tell you that it is actually 5m long. All I am saying is that it is perfectly OK for us to agree to disagree about some aspects of reality.'

The Addition of Velocities

One of the predictions – indeed, absolute requirements - of Special Relativity is that nothing can travel faster than light. Understandably, many people find this difficult to accept; after all, if you get into a rocket and accelerate at a constant rate, surely there must come a point, so the argument goes, when you exceed the speed of light.

There are many ways of explaining this effect but most of them involve ideas about force, mass and energy which we have not yet discussed and which introduce unnecessary complications. The truth is that the effect can be explained using nothing more than the ideas which have so far been introduced in the following way:

Let us suppose that the guard in the train (which you remember is travelling at 60 ms^{-1}) fires a gun forwards down the length of the train at the instant that he passes the station master. The gun is known to have a muzzle velocity of 75 ms^{-1} so, on the face of it, the station master should see the bullet travelling at $60 + 75 = 135$ ms^{-1} which is faster than the speed of light (which, you will recall, is only 100 ms^{-1} in our fanciful scenario).

So that we can do some calculations, let us suppose that, 15 m up the train, the guard has rigged up a target which causes a paintball to be fired onto the platform. Since, to him, the bullet is travelling at 75 ms$^{-1,}$ the time taken for the bullet to reach the target will be $15/75 = 0.2$ s. In the guards frame of reference, the coordinates of the event are (0.2, 15).

Now let us calculate the coordinates of the same event in the station master's frame of reference. We shall use the Lorentz equations quoted on page 11 but we must remember that, since to the guard the platform is moving backwards, we must put $v = -60$. This does not make any difference to the value of k which is still 0.8.

Now $$x' = \frac{x - vt}{k} = \frac{15 + 60 \times 0.2}{0.8} = 33.75$$

so the station master will find the paint 33.75 m up the platform.

The temporal coordinate is calculated as follows:

$$t' = \frac{t - v/c^2 x}{k} = \frac{0.2 + 60/10000 \times 15}{0.8} = 0.3625$$

from which the station master must conclude that the speed of the bullet is 33.75/0.3625 = 93.1 ms⁻¹. In other words, in Relativity, 60 + 75 = 93.1 – not 135!

It is a simple matter to derive a formula for the addition of two velocities by doing the same calculation using letters. In general (remembering that we must put v negative)

$$\frac{x'}{t'} = \frac{(x + vt)/k}{(t + v/c^2 x)/k}$$

if the muzzle velocity of the bullet is u then $x = ut$ so

$$\frac{x'}{t'} = \frac{(ut + vt)/k}{(t + v/c^2 ut)/k} = \frac{u + v}{1 + uv/c^2}$$

It is easy to see that if either u or v is equal to c then the result is c and that it is therefore impossible to achieve a speed faster than c by adding together any number of speeds smaller than c.

So what happens if the train accelerates at a constant rate of, say 5 ms⁻² starting at 60 ms⁻¹?

First we must agree on what we mean by 'accelerate at a constant rate'. Normally this means that every second, the speed increases by 5 ms⁻¹ and the speed of the train would increase linearly – 60, 65, 70, 75, 80 etc. reaching the speed of light in 8 seconds. But if we are taking relativity into account we can't just add 5 ms⁻¹ each time, we must use the above formula for the addition of velocities to work out the speeds every second[14]. This gives us the sequence 60, 63.1, 66.0, 68.75, 71.3 etc. It is clear that, from the point of view of the station master on the platform the train is never going to reach the speed of light.

It should, however, be remembered that, to the guard on the train, the acceleration appears to be smooth and constant. He will feel a steady force accelerating him forward; all he sees is the platform accelerating backwards – but never exceeding the speed of light, of course.

---

14 Strictly speaking we should use the formula every millisecond or even microsecond to get a more accurate result. Better still, we should use a bit of calculus but you get the idea, I hope.

The Penny in the Hole Paradox

'I've had another idea' said Arthur to Betty one day pulling out a large square of plywood with a hole in the middle. 'Look' he said, 'here is a penny a little smaller then the hole and if I drop the penny horizontally like this, it will just fall through the hole. But if the plywood sheet was moving fast enough, the hole would be contracted in the direction of motion and there is no way the penny could pass through the hole. On the other hand, if you were sitting on the sheet you would think that it was the penny which was contracted and the the penny would pass through it easily. So who is right? Will it pass through the hole or not? Either it does or it doesn't.'

'It will pass through the hole,' said Betty confidently.

'But the hole will be too small!'

'No it won't. The penny will be smaller than the hole.'

'So you are saying that your frame of reference is more special than mine? And that Einstein is wrong?'

'No, you have just got to think things through carefully.'

'Go on then, explain it to me,' said Arthur.

'Firstly the situation is a bit complicated by the fact that there are two objects in motion. It will simplify the argument if we assume that you are sitting on the penny and that the sheet is rising slowly upwards while moving rapidly from left to right like this:

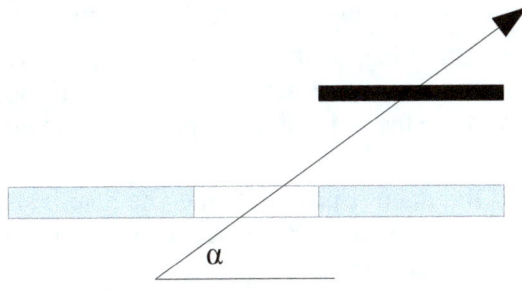

'Yes OK. I see that you have drawn the slot smaller than the hole too. That is good because it shows that the penny will definitely not go through the hole as I said.'

'Yes I have done that deliberately but there is one feature of the diagram which is not correct.'

'What is that?'

'Wait and see,' said Betty mysteriously. 'First let us agree that the the penny and the sheet are very thin so it doesn't matter what the vertical speed is, only that it is small and constant. Suppose this (non-relativistic) speed is u and that the horizontal (relativistic) speed of the sheet is v. This means that the tangent of the angle which I have marked α is equal to u/v.

'Let us also suppose' continued Betty' 'that the diameter of the hole and the penny when stationary is d. So, from your point of view, the hole is contracted to a width kd but I hold the view that it is the penny which is contracted to a width kd (k being less than 1). Are we agreed so far?

'Yes,' Arthur agreed. 'Perfectly.'

'Now let us, for a moment, consider the situation from the viewpoint of a mutual friend, Chris, who is moving to the right with a speed $v/2$. From Chris' point of view, the penny is moving to the left and the sheet is moving to the right both with the same speed $v/2$ and both the penny and the hole are contracted by the same amount. To Chris the penny will fill the hole exactly as it passes through and symmetry dictates that there will be an instant of time when this happens, the left hand edge of the penny will just graze the left hand edge of the hole and similarly with the right hand edge.'

'I don't see what all this has got to do with anything.'

'The important thing,' said Betty, 'is that if Chris sees both edges of the penny come into contact with both edges of the hole, then all other observers must see the same thing happen (though not necessarily at the same time).'

'So what? It remains true that, to me, the hole is smaller than the penny so it stands to reason that, since the penny is horizontal, it can't possibly go through the hole.'

'Who said anything about the penny being horizontal?'

'I did,' said Arthur, 'I said that the penny would be horizontal when it was dropped.'

'Yes – but that was when it was stationary. Now it is moving

44

downwards.'

'What has that got to do with it? I know it is horizontal – after all, I am sitting on the d---d thing!'

'Yes, to you it is horizontal, but to me it is not. From your point of view it is the sheet which is tilted. Look. It works like this:

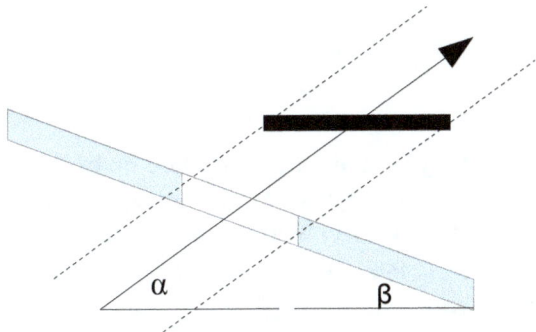

'The sheet is tilted up at an angle β which is such that, as it rises the left hand edge of the hole just grazes the left hand edge of the penny; a short time later, the right hand edge of the hole just grazes the right hand edge of the penny. The penny slides through the hole rather like a letter through a letterbox.'

'But I thought you said that there was an instant in time when the penny exactly fitted the hole.'

'I said that Chris noted that there was such an instant but, to you the left hand edge of the penny enters the hole before the right hand edge. As we shall see in a minute, to me the right hand edge of the penny enters the hole before the left hand edge.'

'But that's absurd; how can we disagree about the order in which two events occur?'

'The crucial point here' said Betty, 'is that the two events occur very close together in time. So close, in fact, that it is impossible to send a signal from one event to the other. Such events are said to be space-like because they are more separated in space than they are in time.'

'It will take a bit of time to get my head round that. So what does the situation look like from your point of view?'

'Like this:' said Betty, showing her brother the following diagram:

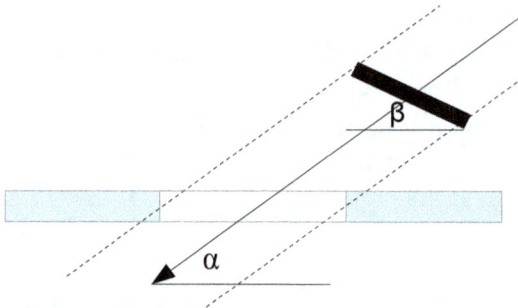

To me, the penny is contracted but tilted up by the same angle β so that the edges of the penny just scrape the edges of the hole, but this time it is the right hand edge which enters the hole first.'

'Well I never!' exclaimed Arthur. 'You just can't trip this fellow Einstein up can you?!

'Can we work out the angle β?' he asked.

'Yes. Easily. It doesn't matter which point of view we adopt, the answer will be the same. If we look in detail at the instant at which, to me, the penny just enters the hole, the situation will look like this:

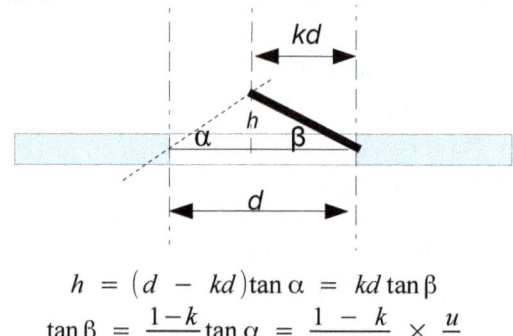

$$h = (d - kd)\tan \alpha = kd \tan \beta$$

so

$$\tan \beta = \frac{1-k}{k} \tan \alpha = \frac{1-k}{k} \times \frac{u}{v}$$

'Well, that's pretty cool! I would never have thought that things could get tilted by their vertical motion.'

'But they do.' said Betty.

'Wait a minute. I have just thought of something. We agreed at the start that the penny was horizontal before it dropped. How can it suddenly tilt upwards when it starts to move?'

'You have put your finger on an excellent point there, and the answer is best explained using a related paradox which I call the Carrom board paradox. Would you like to hear about it?'

'Go on then,' said Arthur.

'Suppose that you are playing Carrom or some such game where a disc, sliding along a table falls through a hole in the table.'

'I get it! From the point of view of the people playing the game, the penny is shrunk and it will easily fall through the hole; but to an ant sitting on the penny the hole is shrunk and there is no way it can fall through.'

'Correct. And can you see how the paradox is resolved?'

'Well, no, I can't. The disc is sliding along the table so it cannot possibly tilt – not until it reaches the edge of the hole at any rate. And if the hole is shrunk to less than half the width of the penny, the front of the penny will reach the other side of the hole before the centre of gravity of the penny has reached the edge.'

'And yet the penny does fall through the hole.'

'I don't see how it can.'

'Suppose, for the sake of argument,' said Betty 'that in the table frame the penny (which is shrunk to half the width of the hole) stays horizontal the whole time but at the instant that the rear of the penny passes the edge of the hole – i.e. at the moment that the penny becomes unsupported – it immediately acquires a small vertical downward velocity.'

'OK. But I have just pointed out that at this instant in the penny's frame, the penny is bridging the gap.'

'And there is your error.'

'What do you mean?'

'In the table frame there is an instant at which the whole penny acquires a vertical velocity – but events which are simultaneous in one frame are not simultaneous in another. In the penny frame, the front of the penny starts to move downwards *before* the back of the penny has got to the edge of the hole like this:'

'You mean the penny *bends*?' said Arthur incredulously.

'Exactly that.'

'But surely that is absurd. A penny can tilt but it can't *bend*.'

'Well according to the ant, it doesn't bend – but to the people playing the game it does. It is just another of those apparent distortions like length contraction which affect things in motion.'

'But if the penny was made of some brittle material like glass it would break!'

'Possibly.'

'But from the point of view of the players, the penny is simply shrunk and it will just fall through the hole without breaking. That is a clear contradiction.'

'Perhaps not.'

'Why not?'

'If we are not careful we are a liable to confuse here what happens in an ideal world with what might or might not happen in the real world. Firstly, the penny is not *bent* it is *sheared*. Secondly we must assume that the penny has a finite thickness (otherwise there is nothing to shear). Thirdly we must assume that the vertical acceleration is very large but finite. In this case, the shearing will be gradual, not sudden, and will progress from right to left along the penny.

'Now if we are to apply the usual laws of elasticity to this object, it is obvious that we can only apply them with consistency in the frame in which the object is at rest. (Nobody would suggest that a space ship travelling at a constant relativistic speed is under any actual strain as a result of its length contraction).

'Now in the ant's frame, the whole penny suddenly begins to accelerate to a relativistic speed while staying horizontal the whole time. When the penny has acquired a relativistic speed in the vertical direction it will be length contracted in thickness. Some authors argue that this contraction is merely a change in coordinates and will have no physical consequences. I, myself, believe that, while a *constant* length contraction has no physical consequence, a *changing* length contraction does and that the acceleration will set up vertical strains in the penny which may be sufficient to cause it to break.'

'I can't say that I am entirely happy with that explanation, but I suppose it will have to do for the moment,' said Arthur, 'after all, I agree that the whole situation is pretty unrealistic anyway.'

Author's note: The whole question of whether the ordinary rules of physics can be applied to accelerated frames of reference is a minefield and there is still much disagreement as to the correct way of applying the rules. Further discussion of this issue will be found in the chapters on 'The Broken Rope' and 'The Space Station'.

Tħe Past and Future Present

The pole in the barn paradox and the paradoxes of the penny and the hole are puzzling because we find it very difficult to accept Betty's statement that 'events which are simultaneous in one frame are not simultaneous in another'. There seems to be some fundamental illogicality here which goes far beyond rulers shrinking or clocks running slowly. If in one frame of reference A precedes B how can there be another frame in which B precedes A? Indeed – if A was the *cause* of B then surely B cannot also *cause* A?

It is important therefore to sort out what kinds of events can be reordered and which cannot. Lets simplify things by concentrating on motion in one direction only and plotting a graph of the motion of an object – a space ship, say – by plotting its position on the horizontal axis and time vertically – like this:

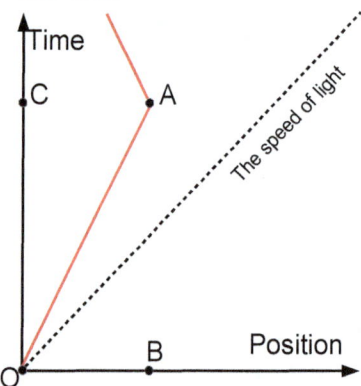

The red line could represent Albert's journey to a Alpha Centauri at a constant speed of half the speed of light. A line like this is called a 'world-line'.

Four 'events' (points in space and time) are marked on the graph. O is the origin (0, 0)[15] and represents the point when Albert sets off. A is the point in space and time when he reaches his destination. Since Alpha Centauri is 4 light years away, he will get there in 8 years and according

---

15 In line with usual practice we shall always put the temporal component first. Note, however, that this is slightly at odds with the normal convention of placing the X coordinate first.

to Ludvig at any rate, the coordinates of A are (8, 4). (We shall not be concerned with Albert's view of the situation here.)

B represents the coordinates of the star at the time Albert sets out (0, 4) and C represents the coordinates of his stay-at-home brother Ludvig at the instant Albert reaches the star. (8, 0) In diagrams like this we usually arrange for the speed of light to be represented by a 45° line and if we work in years and light years, we can talk the speed of light being equal to 1.

Now it happens that, at the instant Albert leaves Earth, a mutual friend Klara passes Earth in a fast spaceship travelling in the same direction so at this instant, her coordinates are also (0, 0). She is travelling at 60% of the speed of light so the relativistic factor $k = \sqrt{1 - 0.6^2} = 0.8$ and in her frame of reference, the coordinates of Albert's arrival at Alpha Centauri are:

$$x' = \frac{x - vt}{k} = \frac{4 - 0.6 \times 8}{0.8} = -1 \text{ light years}$$

$$t' = \frac{t - v/c^2 x}{k} = \frac{8 - 0.6 \times 4}{0.8} = 7 \text{ years}$$

(Note that the position coordinate is negative because she is travelling faster than him and has left him behind.)

Every event in Ludvig's frame has a corresponding coordinate in Klara's frame. We can visualise the mapping between the two frames by superimposing a grid representing Klara's coordinate system (in black) on top of Ludvig's (in brown) as shown below.

Notice how Klara's spatial axis (labelled Klara's 'now') has been tilted anticlockwise and the temporal axis (labelled Klara's 'here') clockwise by the same degree – like a pair of scissors. The point (8, 4) in Ludvig's frame has become the point (7, −1) in Klara's. Note also how points along the red line have transformed into points along the red line. (e.g. (4, 4) in Ludvig's frame has transformed into (2, 2) in Klara's frame). This is because, of course, both Ludvig and Klara will calculate the same value for the speed of light.

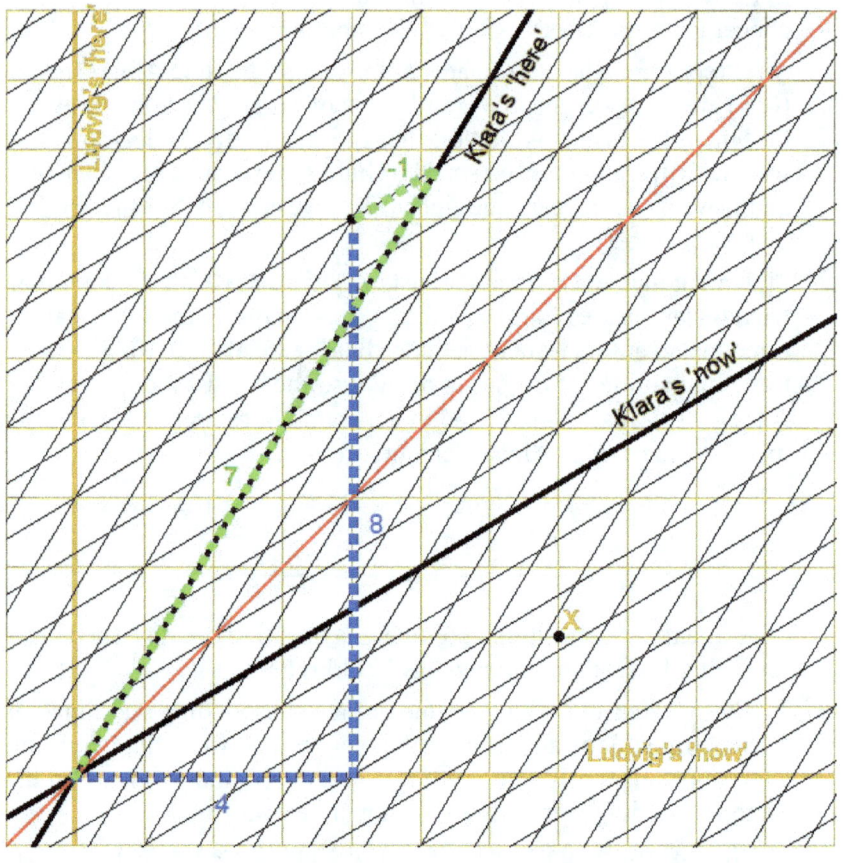

Klara's spatial axis contains all the points with coordinates like (0, *x*). In other words this line represents all those events which, to her, are happening 'now'. Of course, Ludvig's 'now' is the horizontal axis. What this means is that, even though at time *t* = 0 Klara and Ludvig are in the same place (Earth), they must differ as to which events happening elsewhere in the universe are happening at the same time.

If Klara were to move at the speed of light, her spatial and temporal axes would coincide along the red line but as long as her speed is less than the speed of light, she can make her 'now' line point anywhere inside a cone which is defined by the two red lines shown opposite.

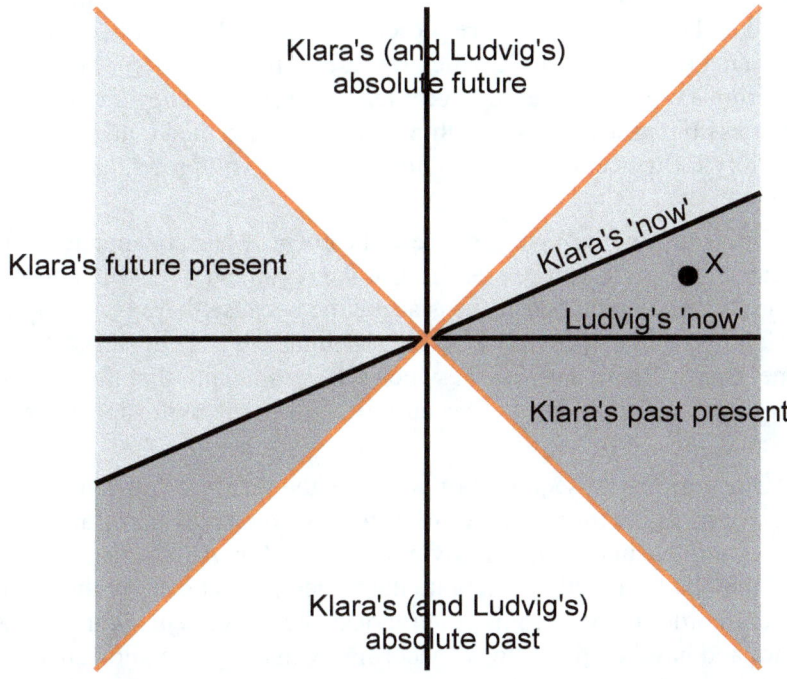

Remember, at the instant depicted in the diagram, both Klara and Ludvig are in the same place (at the origin) but Klara is moving rapidly to the right. By altering her speed Klara can make her 'now' line point anywhere within the grey areas. She cannot, however, make her 'now' line cross into the white areas. Events which occur in the white are above her 'now' line are in her (and Ludvig's) absolute future. These are the events which they can expect to influence. Events which took place in the white area below the 'now' line occurred in her (and Ludvig's) absolute past. These are the events which could, potentially, have influenced things happening at the origin.

The grey areas contain all those events which cannot affect or be affected by something which happens at the origin because to get from one event to the other, a causal influence would have to travel faster than light. Since these events are neither in the future nor in the past, they are, in a sense, all part of Klara's and Ludvig's 'present'. Events in the light grey area are all above Klara's 'now' line and are therefore in her potential future. I like to call these events Klara's 'inaccessible future' because Klara cannot do anything about them. Alternatively we could call them Klara's 'future present'. Similarly the events in the dark grey area (like the event X which represents the explosion of a supernova) appear to her to have already happened. They are in her 'inaccessible past' (because nothing that happened then could have possibly influenced her present situation) or, if you prefer, her 'past present'.

But from Ludvig's point of view, the event X hasn't happened yet because it is in his future present. But the argument over whether event X has or has not happened yet is academic because there is no way in which Ludvig can possibly stop it happening (he can't even get there in time!), nor is there any way in which Klara can claim that the event can cause any change in her life because light from the explosion has not yet reached her.

One curious consequence of all this is the fact that you can apparently make time go forward or backward on a distant star at will. For example, suppose that Ludvig has a friend on Alpha Centauri and that they have agreed to celebrate the millennium at exactly the same time. In order to synchronise their clocks they have sent signals to each other and have set their clocks accordingly making due allowance for the fact the the signals take 4 years to travel from one to the other. Each are now agreed that the millennium will take place on Earth at the coordinate (0,0) and on Alpha Centauri at (0, 4). Note that, since there is no relative motion between them, they share the same reference frame.

In Klara's frame of reference, however, the coordinates of the millennium on Alpha Centauri are:

$$x' = \frac{x - vt}{k} = \frac{4 - 0}{0.8} = 5 \text{ light years}$$

$$t' = \frac{t - v/c^2 x}{k} = \frac{0 - 0.6 \times 4}{0.8} = -3 \text{ years}$$

In other words, according to Klara, the millennium on Alpha Centauri happened 3 years *before* the millennium on Earth! (when she was still 5 light years away from that star). Because Klara's 'now' line has been tilted upwards, events occurring on stars in front of her seem to be shifted backwards in time (and events behind her, forwards) by an amount which is equal to 3 years for every 5 light years.

In general, we can say that, if you move towards a star which is a distance d in front of you (as measured in the frame in which the star is stationary), time will appear to jump forwards (causing events to appear to have happened earlier) by an amount equal to dv/c^2.

Now the Earth is rotating round the sun at a speed of about $0.0001c$. What this means is that, when we are at that time of year when we are approaching a distant galaxy like the Andromeda galaxy which is 2.5 million light years away, time on the galaxy advances by $2,500,000 \times 0.0001 = 250$ years! 6 months later, time will have slipped back the same amount. Just *walking* towards a really distant galaxy will change the time on the galaxy significantly – from your point of view that is!

Armed with this new idea, it is possible to explain the Twin's Paradox in a completely new way. You may remember that Albert takes 3 years to reach his destination, Alpha Centauri. During the journey he argues that his brother's clocks should be running slow because of their relative motion by a factor 0.6 and that when he reaches the star, his brother's clock should read $3 \times 0.6 = 1.8$ years. The same is true on the return journey so he might reasonably expect his brother to be only 3.6 years older when he gets back home. What this analysis leaves out, however, is the instantaneous change in Albert's 'now' when he turns round. Since his speed changes from -0.8 to $+0.8c$, time back on Earth will jump forward by $2 \times 4 \times 0.8 / 1^2 = 6.4$ years which, together with the 3.6 years already accounted for, adds up to 10!

Now you may be thinking that this is completely absurd. How can Albert change the time on Earth by just by turning round reversing his direction of motion? But the question is misleading. Suppose you start at a tree and walk 100 m away from it. The tree is now 100 m behind you. Turn round. Suddenly the tree is 100 m in front of you. It is nonsense to ask 'how can you move a tree which weighs several tons and is rooted to the ground, a distance of 200 m in the blink of an eye?' All that has happened is that you have changed your frame of reference. It is the same with Albert. All that has happened is that he has stepped off a frame of reference which is moving away from Earth onto a frame of reference which is moving towards Earth.

Interval

No, this is not the point where I hand out ice creams! The '*interval*' between two events has a very specific meaning in Special Relativity.

In a non relativistic world, two events are separated both in distance and in time and two observers in relative motion will always agree on both of these values. They are absolute. But in Special Relativity, space and time get a bit mixed up and two similar observers will not agree on either the distance between the events or the time interval between them (or even, as we have seen, in some cases, the order in which the two events occur.)

There is, however, a simple quantity which they can agree upon which is the quantity

$$I = \sqrt{t^2 - x^2/c^2}$$

which would have the dimensions of time. (We are assuming here a 'one dimensional' universe in which y and z are zero.)

For example, we have seen how in Ludvig's frame, Albert arrives at his destination at (8, 4) but in Klara's frame his coordinates on arrival are (7, −1). According to Ludvig, therefore, the *interval* between his leaving Earth and arriving at Alpha Cantauri is

$\sqrt{8^2 - 4^2} = \sqrt{64 - 16} = \sqrt{48} = 6.93$ (Remember, we are working in years and light years so $c = 1$.)

Klara's calculation goes like this:

$\sqrt{7^2 - (-1)^2} = \sqrt{49 - 1} = \sqrt{48} = 6.93$ which is exactly the same.

Lets calculate the interval between the origin and the explosion of the supernova at X. Let us suppose that in Ludvig's frame the explosion occurs 7 light years away and 2 years in the future (see the diagram on page 53). Ludvig's calculation goes:

$I = \sqrt{2^2 - 7^2} = \sqrt{4 - 49} = \sqrt{-45} = 6.71\,i$. (The imaginary answer merely indicates that the interval is space-like.)

In Klara's frame the event occurs 7.25 light years away and 2.75 years in the past. Her calculation goes as follows:

$I = \sqrt{2.75^2 - 7.25^2} = \sqrt{7.56 - 52.6} = \sqrt{-45} = 6.71\,i$

In ordinary Euclidean space, a map of all the points equidistant from the origin is, of course, a series of concentric circles. But in spacetime, a map of events at equal intervals from the origin form a series of hyperbolae with some regions being an imaginary (time-like) interval away from the origin.

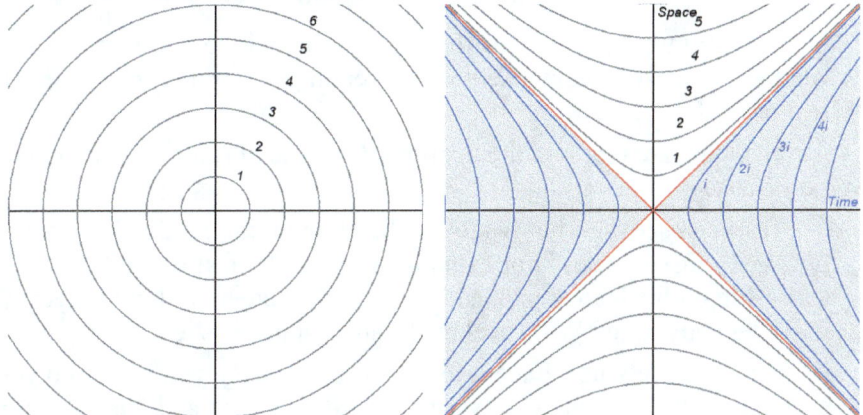

In three dimensions we define the 'interval' between any event (x, y, z, t) and the origin to be

$$I = \sqrt{t^2 - (x^2 + y^2 + z^2)/c^2}$$ [16]

It is, of course, the fact that that the spatial components are subtracted instead of added which makes all the difference between the behaviour of distances in Euclidean space and intervals in spacetime.

---

[16] You could equally well make the spatial components positive and the temporal component negative (and many authors do) but this makes the interval between events which occur approximately in the same place (a condition which is true of virtually all common events) imaginary, not real. In addition, I have chosen to divide the expression by c so that interval has the units and dimensions of time. Some authors even leave the c^2 factor out altogether. This may make sense to a mathematician who is quite prepared to say that $c = 1$ and pretend that distances and times are no different, but it is anathema to a physicist like myself.

The Broken Rope

or Bell's Spaceship Paradox

'I am still troubled,' said Arthur one day 'about the question of whether all these effects, like pennies bending and trains contracting, are real or not.'

'Yes, they are real – in the appropriate frame of reference at any rate,' said Betty.

'Yes I can see that; but would the bending or contraction have any *physical* consequences? For example, suppose I am sitting in a spaceship on the launch pad, and then I accelerate up to light speed – will I feel anything? Will I feel as if I am being squashed?'

'No, absolutely not. Except for the force pressing you back into your seat due to the acceleration, everything will just look and feel perfectly normal.'

'But from your point of view, you see me becoming more and more squashed.'

'Yes, but that's only of academic interest really and due to the way light behaves when I try to measure the length of your space ship.'

'So length contraction is just an illusion.'

'No, it's not just an illusion. From my point of view, your space ship really does get shorter. Look, suppose you tied two identical spaceships together with a long rope and arranged for the two spaceships to start accelerating at the same instant.'

'How would you do that?' asked Arthur.

'Well suppose Albert and Ludvig were piloting identical space ships and that Klara, positioned exactly halfway between them, flashed a light. The light would reach them both at the same time and the race would begin.'

'I see what you are getting at. The two spaceships are always travelling at exactly the same speed so the distance between them will always remain exactly the same and the rope will not break.'

'Right. But when they got up to a sizeable fraction of the speed of light, the rope would be length contracted in Klara's frame and it would have to break.'

'I don't buy that. Surely the distance between the two space ships would also be length contracted so the rope would stay intact.'

'You have put your finger on a very subtle point there. When we talk of distances being length contracted are we talking about objects like spaceships and rulers getting shorter *within space* or are we talking about *space itself* getting smaller. Or are we talking about something else entirely?'

'I don't know. You tell me.'

'Let's start by simplifying things so that we don't have to worry about things like acceleration. Let's suppose that at the instant the two brothers see the light signal, they instantly start moving at a fast speed – say 60% of the speed of light. Let us also suppose that they are initially 1000 m apart.'

'OK. How about it if we plotted one of those time/space diagrams?' suggested Arthur.

'Excellent idea,' said Betty.

As usual, Time is plotted on the vertical axis and position on the horizontal axis. The lines A and L are Albert's and Ludvig's world lines as seen by Klara and at time $t = 0$ Albert and Ludvig start moving.

The red 45° lines represent potential light rays. The thick horizontal blue line represents the distance between the space ships – and hence the length of the rope – before the ships start to move.

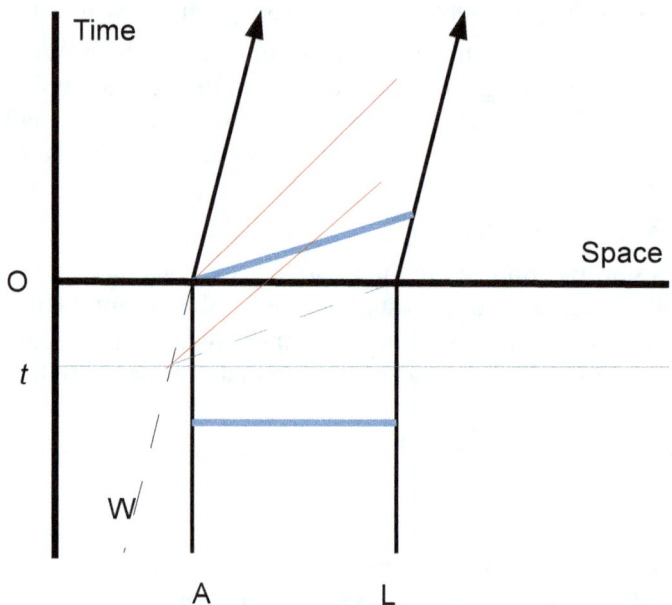

'Now,' continued Betty, 'let's consider what the situation looks like from the point of view of someone who is travelling at just the right speed to catch up with Albert at the instant he starts to move. Let's call him Walter.

'From Walter's point of view, Albert and Ludvig are travelling towards him at 60% of the speed of light.'

'But won't he see them contracted?'

'Yes – excellent point. To Walter they will only be 800 m apart and the rope will only be 800 m long.[17]'

'There is something else I have realized.' said Arthur. 'Although in Klara's frame, Albert and Ludvig set off at exactly the same time, this won't be true in Walter's frame, will it?'

'Another excellent point!'

'From Walter's point of view, the light pulse which is travelling towards Ludvig will reach him first because he is travelling towards the oncoming light; whereas the light travelling towards Albert has got to

---

17 At 60% of the speed of light $\quad k = \sqrt{1 - 0.6^2} = 0.8$

chase him down the line!'

'Spot on! Now as we know, his *now* line is inclined upwards so he sees Ludvig starting to move at the *earlier* time t and in his frame, by the time he has reached Albert, Ludvig has already travelled some distance and the distance between the two spaceships (shown by the inclined blue line) is increased.[18] There is therefore not a shadow of doubt that, in Walter's frame of reference, the rope is going to break.'

'So it must break in Klara's frame too! That proves that length contraction is not just some trick of perspective like the foreshortening of a table when looked at obliquely, it is a real effect which can have real physical consequences!'

'Yes, the rope will definitely break.'

'Let me get this absolutely clear. In Klara's frame, both ships move off at exactly the same speed at the same time and remain the same distance apart. So it is not *space* which gets contracted only *objects within space* – is that right?'

'No, not really. As we agreed, from the rope's point of view, it stays the same length all the time; it is the space ships which move further apart.'

'But you have just said that the space ships remain the same distance apart!'

'In Klara's frame, yes.'

'I am confused!'

'You are confused because you are still thinking of space as some sort of object which either can or cannot be contracted. You must try to get rid of this notion altogether. All that exists in space are *coordinates*. In Klara's frame the X coordinates of the two space ships increase linearly with time at the same rate. The difference between these coordinates (i.e. the separation of the two space ships in her frame) remains constant. The coordinates of the ends of the rope do not, however, accelerate at the same rate and so the length of the rope gets shorter.'

'That doesn't make sense. If the ends of the rope are attached to the space ships they have to accelerate at the same rate.'

---

18 It is actually the *component* of this line along the X axis which is increased.

'Which is why the rope has to break.'

'But suppose the rope is only attached to Ludvig's ship. What then?' asked Arthur.

'We are in great danger again of trying to apply the laws of physics to an idealised situation which is wholly impossible in practice, but let's make the scenario a bit more plausible by supposing that the two ships accelerate at a finite rate of, say, 1G. We must also make some assumptions about the nature of the rope. If we assume that the rope is inextensible (another impossibility) then what we are saying is that at all times, *in the inertial frame in which the rope is instantaneously at rest*, the distance between the end points is 1000 m. If that is the case then, as the rope gets faster and faster, in Klara's frame it gets shorter and shorter. And since the front end of the rope is accelerating at a constant rate of 1G, we must conclude that, in Klara's frame, the rear end of the rope is accelerating faster than the front end. In other words it is going to part from Albert's ship.'

'What will Albert see then?'

'Well, since both ships start accelerating from zero, Albert will see[19] his brother start off at the same time but although Klara sees both ships accelerating at the same rate, Albert will see Ludvig accelerating more quickly than him. However you look at it, the rope is definitely going to break.'

'Well, I suppose I am 90% convinced,' said Arthur doubtfully.

'Think about it from Walter's perspective,' concluded Betty. 'There is absolutely no doubt about it from his point of view – the rope is definitely going to break. And although Albert and Ludvig disagree about the reason why the rope breaks, they will both agree that it does.'

---

19 Once he has taken into account the finite speed of light, of course.

The Red Danger Signal

The accident could have been much worse. The new High Speed Train was only doing half its maximum speed when it hit the facing points at the end of the platform and, while it derailed, it stayed upright and the driver and the guard – the only people on the test train at the time – were uninjured. The driver insisted that the home signal was green – a claim vigorously supported by the guard – but the signalman and the station master were adamant that the signal was showing red and that the driver should have stopped.

Later that day the station master and the guard were discussing the accident over a pint.

'I just can't understand it,' said the guard, ' I *know* I saw a green light. I swear it.'

'Yes.' replied his friend, 'But I have had an idea about how it might have happened.'

'How?'

'You know when an ambulance goes past its siren seems to dip in pitch?'

'Yes, it's called the Doppler effect.'

'Well, I looked it up in one of my old Physics books this afternoon and apparently there are two Doppler effects – the Moving Source effect and the Moving Observer effect – with slightly different formulas[20]. I think that one of these might solve the riddle.'

'Of the two I would guess it was the Moving Observer effect.' replied his friend, 'because it was the train that was moving, not the signal. What was the formula?'

'The book said that the wavelength of the light would appear to be shortened by a factor $\dfrac{c}{c + v}$ where c is the speed of sound and v is the speed of the moving observer.'

'What happens if you put the numbers in, assuming that the Doppler shift works in light as well as sound? I know that the train was travelling at 28 ms^{-1} when it entered the station because I was monitoring the

---

20 See the Appendix for a further explanation of these effects.

speed of the train the whole time.' said the guard.

'Because the speed of light is 100 ms^{-1} ' said his friend, pulling out a calculator, ' the shortening factor works out to be 100/128 = 0.78.'

'But do you know the wavelengths of red and green light?'

'Yes, I looked those up too. The wavelength of red light is 800 nm and the wavelength of green light is 600 nm.'

'Well, there you are then. That solves it. The red light was Doppler shifted into the green and that's why the driver and I thought the signal was green while you though it was red.'

'But there is a slight problem here' said the station master. '800 × 0.78 is 625, not 600.'

'Surely that's near enough isn't it?'

'No. If the light which you saw had a wavelength of 625 nm it would have looked yellow, not green.'

'Well it definitely was green, I can tell you. By the way – it's your round I think.'

While the station master was busy at the bar, the guard had a thought of his own and when his friend returned with the beers he burst out with:

'I know what's wrong! We are using the wrong formula! From the driver's point of view it is not the train which is moving, it is the signal – so we should be using the Moving Source formula, not the Moving Observer one.'

'That's a really good point. Lets see if it checks out. If the source is moving towards the observer the book says that the wavelength should be shortened by a factor $\dfrac{c - v}{c}$ which is 72/100 or 0.72.'

'That looks promising. It will be shortened more. What does the wavelength work out to be?'

'800 × 0.72 is 576.'

'Bummer. That's too short isn't it?'

'Yes. A light of that wavelength would look distinctly blue.'

'Well I don't know what the answer is. Let's talk about something else.'

So for a while the two friends chatted amiably and, inevitably, consumed a few more beers. Again, it was while his friend was buying yet another round that the guard had another idea and when the beers arrived he said:

'You remember when we were working out why the paint was spilled in the wrong place on the train, we decided it was because when objects were moving, lengths were contracted. Perhaps that's why the light looked green. It was nothing to do with the Doppler shift, it was relativistic contraction.'

'Well, there is an easy way to find out if that idea is right. Lets work out the figures. The formula for the Length Contraction k is

$$\sqrt{1 - v^2/c^2} \text{ which equals } \sqrt{1 - 28^2/100^2} = 0.96 \text{ .'}$$

'That's no good.'

'You're right. The wavelength works out to be 768 nm which is only slightly orange.'

'Wait a minute, though' said the guard, 'what if the wavelength is both Doppler shifted *and* contracted?'

'That's not going to work. From the driver's point of view, the wavelength is already Doppler shifted to a wavelength of 576 nm. If it is contracted by a further 0.96 it will have a wavelength of 553 which is definitely in the blue region.'

'Well, what if we use the Moving Observer effect instead of the Moving Source formula?'

'Surprisingly, that seems to work. 625 × 0.96 is exactly 600 nm!'

'That's really strange. But we must be on to something here. Actually, I am not sure we can just apply the length contraction formula here anyway. That applies to things like trains and platforms, not to wavelengths.'

'So where does relativity come in?' asked the station master.

'I am not sure. How about this for an idea. You know we also found out that moving clocks run slow so from the driver's point of view, the atoms in the red light will be oscillating more slowly than they should...'

'But that means that, to the driver, the wavelength should look *longer* than 600 nm which would put it in the infra red and make it completely invisible!'

'Yes, but I am assuming that the Moving Source Doppler effect will still apply.

'So what you are saying is that, to the driver, the wavelength is *lengthened* by a factor 1/0.96 and *shortened* by a factor of 0.72. Is that it?'

'Absolutely. How do the figures work out?'

'Well 800 / 0.96 is 833.3 and 833.3 × 0.72 is 600!'

'That's it! The effect is a product of Time Dilation and the Moving Source effect.'

'But it seems strange that we got the same result when we used the Moving Observer effect and Length Contraction.'

'Well, let's see what we get if we use letters instead of numbers. Overall the wavelength will be shortened by a factor equal to

$$\frac{1}{k} \times \frac{c - v}{c} = \frac{c - v}{c\sqrt{1 - v^2/c^2}} = \frac{c - v}{\sqrt{c^2 - v^2}}$$

'But $c^2 - v^2$ can be factorised as $(c - v)(c + v)$ so the shortening factor will be

$$\frac{c - v}{\sqrt{(c - v)(c + v)}} = \sqrt{\frac{c - v}{c + v}} \; ,$$

'What a pretty formula! It seems to strike a very happy medium between the Moving Observer and the Moving Source effects' commented the station master, 'and the numbers check out too because

$$800 \times \sqrt{\frac{100 - 28}{100 + 28}} \text{ is exactly 600!}$$

'Of course, if we used the Moving Observer formula and Length Contraction we would get a shortening factor of

$$k \times \frac{c}{c + v} = \frac{c\sqrt{1 - v^2/c^2}}{c + v} = \frac{\sqrt{c^2 - v^2}}{c + v}$$

which is also equal to $\sqrt{\frac{c - v}{c + v}}$. But I still think it is better to regard the effect as due to Time Dilation and the Moving Source effect because it is only the driver who sees the change in colour and, to him, he is always stationary.'

The guard is correct. The driver is, by definition stationary and so there is no Moving Observer effect in light. On the other hand, as the source of light movers towards him, the wavelengths are all bunched up in accordance with the Moving Source effect. In addition, however, in his frame of reference the atoms in the signal are vibrating more slowly than they would if the signal was stationary which would increase the wavelength. The combination of the two effects is the Doppler effect in light.

But having said that, there is nothing really the matter with the Moving Observer plus Length Contraction alternative – as always, it just depends on your point of view.

Part 2: All about Appearances

The Elastic Train

'I am fascinated by the phenomenon of length contraction,' said Arthur one day. 'I would love to be able to take a photograph of, say, a train squashed to half its length!'

'Yes – but it is worth considering exactly what the photograph would show.' replied his sister.

'What do you mean?'

'Well, suppose the train is 100 m long (when stationary) and that light travels at 100 ms^{-1}. If the train travels at 28 ms^{-1} then $k = \sqrt{(1 - 0.28^2)} = 0.96$. From your point of view, therefore, the train is only 96 m long because of length contraction. Let us suppose that you are standing beside the track and you take your photo well before the train reaches you. When you examine the photo you notice that the rear of the train has just emerged from a bridge which you know is 200 m away. Where do you think the photo will show the front of the train?'

'Well, obviously, since the train is (to me) only 96 m long the front of the train will be only 104 m away.'

'I am afraid you are wrong.'

'How so?'

'How long will it take the light from the rear of the train to reach you?'

'200 m at 100 ms^{-1} that is 2 s.' answered Arthur.

'Correct. And how far will the train move in that time?'

'2 times 28 – 56 m. Hey! That means that the front of the train will only be 200 – 56 – 96 = 48 m away. Is that right?'

'No, not quite. You have forgotten to take into account the time it takes for the light to get from the front of the train to the camera. Think of it like this. There is a race going on between the train and the light. From your point of view the light from the rear of the train is overtaking the train at a speed of 100 – 28 = 72 ms^{-1}. It takes 96/72 = 1.33 s to do this at which point it joins light from the front of the train and reaches you in a further 0.33 s. In the 1.33s it took for the light to overtake the train time the train travelled 1.33×28 = 37.33 m. So all we have to do to find the apparent length of the train is to add this to the length of the train. This gives us 133.33 m and on the photo the front of the train will be 66.67 m away..

'But that means the train will look *longer* not shorter! I was hoping to see it squashed.'

'No, it *is* squashed – it just *looks* longer.'

'What is the difference? If relativity says it *is* squashed, why doesn't it *look* squashed?'

The effect we are considering here has nothing to do with relativity at all. The train looks longer simply because it takes longer for light to reach you from the back of the train than from the front, and in that extra time the train has moved forwards.

'Let's say you turn round and take a photo of the train as it moves away from you,' continued Betty. 'Now the light from the front of the train only takes 0.75 s to travel the 96 m along the train because the relative speed between the train and the light is 128 ms^{-1}. In this time the train moves away from you a distance of 0.75×28 = 21 m so the train will *look* as if it is only 96 – 21 = 75 m long. Is that squashed enough for you?'

'I suppose. But I have thought of something else; suppose I take a *cine* film of the train passing. Surely it can't suddenly appear to shrink from 133 m to 75 m between one frame and the next?'

'No, of course not. It shrinks gradually as it passes you.'

'How can that be?'

'Well, consider the front of the train when it is 100 m away. Light from the train takes 1 s to reach you. Now consider the train when it is 90 m away. At 28 ms^{-1} it takes 10/28 = 0.357 s to cover the 10 m. Light from the front of the train now takes 0.90 s to reach you so it looks to you as if the train has covered 10m in 0.9 + 0.357 – 1 = 0.257 s. This means that the apparent speed of the train is 10/0.257 = 38.8 ms^{-1}! In other words, when an object is moving towards you, it appears to be going faster than it really is. (Remember, this is nothing to do with relativity – it is basically just the standard Doppler shift in action.)'

'What if the train is moving away?' asked Arthur.

'Consider what happens when the train moves from 90 m to 100 m. The extra time taken by the light to get back to you is still 0.357 s but the time interval between the arrival times is now $1 + 0.357 - 0.9 = 0.457$ s and the train appears to be going at a speed of $10/0.457 = 21.9$ ms^{-1}. Quite a lot slower.'

'So on my cine film it will show the train apparently speeding towards me at 38.8 ms^{-1}; but when the front of the train passes me it will suddenly slow down to 21.9 ms^{-1}; the rear of the train still looks as if it is travelling at the faster speed so the train will appear to shorten. Is that it?'

'Absolutely right. Well done.'

'Does this mean that when a radar speed trap measures the speed of an approaching can, it overestimates the speed?'

'No it does not. The reflected radar is Doppler-shifted up in frequency and the change in frequency is used to calculate the actual speed of the car, not its apparent speed.'

'Pity.'

'Let me just state once again, this has nothing to do with relativity. It is just a consequence of the finite speed of light. The formula for the apparent speed of an object moving towards you at a speed v is $v\dfrac{c}{c-v}$ and away from you is $v\dfrac{c}{c+v}$.'

'That's really cool. Hey – wait a minute,' said Arthur, 'there is something weird here. Suppose the train is travelling at half the speed of light i.e. $v = c/2$. If I plug this into your formula for the approaching train it looks as if the train is travelling at the speed of light!. In fact at speed greater than $c/2$ it will look as if it is travelling faster than light! Is that possible?'

'Yes. It only *looks* as if it is travelling that fast. Unlike length contraction which is a real effect, the change in the apparent speed of the train really is just an illusion.'

'What is the formula for the apparent length of the train?' asked Arthur.

If the train moving towards you it will appear to be $l\dfrac{c}{c-v}$ and away from you $l\dfrac{c}{c + v}$ where l is the contracted length of the train.

Something rather interesting happens if we factor in the relativistic length contraction as well; the two formulae become

$$l\sqrt{1 - v^2/c^2} \times \dfrac{c}{c \pm v}$$

which simplifies to:

$$l\sqrt{\dfrac{c \pm v}{c \mp v}}$$

Putting in the figures we used earlier for a train 100 m long travelling at 28 ms^{-1} it will appear to be $100\sqrt{\dfrac{100 + 28}{100 - 28}} = 133.3$ m long when approaching and $100\sqrt{\dfrac{100 - 28}{100 + 28}} = 75$ m long when receding.

Doppler Dilation

We are all familiar with the Doppler shift in frequency as an ambulance goes past. The effect Betty and Arthur have been discussing in the last chapter is exactly the same effect but as it affects distances rather than times. Of course, we do not normally experience the Doppler shift of distance in everyday life because we do not usually use pulses of sound to measure distance but an ultrasonic tape measure does. It works by sending out a a pulse of sound and timing the return signal. If you just measure one pulse, the tape measure will correctly calculate the distance to the wall at the instant the sound bounced off the wall, regardless of the motion of the wall. But if you attempt to measure the length of a flat wagon with reflecting boards mounted front and back by measuring the time interval *between* the reflected pulses, then you will get a Doppler-dilated answer.

In order to prove the formulae quoted in the last chapter it is immensely useful to draw a distance/time graph.

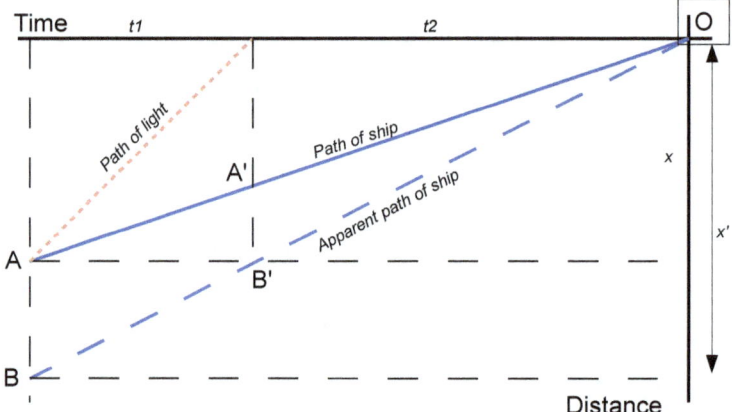

A space ship is approaching the origin O at a speed *v* as represented by the solid blue line. The ship reaches the observer at time *t* = 0. At the point A, a distance x from the observer) it emits a flash of light which takes a time t_1 to reach the observer. In this time the space ship has moved to the point A', but to the observer he sees it as if it was at B'. After a further time t_2 the ship has reached the observer so, to the observer, it looks as if the ship has moved a distance *x* in a shorter time t_2 i.e. it looks as if is is moving faster than it actually is and that when the light was emitted, the space ship appeared to be at B not A.

Once we have got the diagram sorted out it is easy to do the algebra.

$$t_1 = \frac{x}{c} \text{ and } (t_1 + t_2) = \frac{x}{v} \text{ so } t_2 = \frac{x}{v} - \frac{x}{c}$$

$$\text{apparent velocity } v' = \frac{x}{t2} = v\frac{c}{c - v}$$

$$x' = v'(t_1 + t_2) = x\frac{v'}{v} = \frac{xc}{c - v}$$

But what causes the apparent sudden change of speed as the space ship passes the observer? Lets draw the continuation of the diagram:

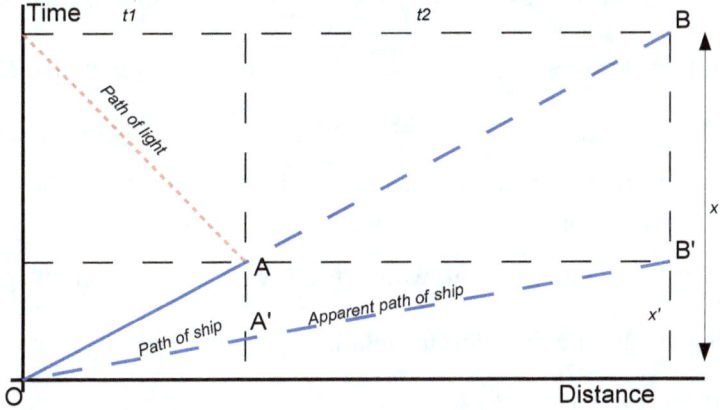

Now when the ship emits a flash of light it is travelling backwards so by the time it reaches the observer the ship has reached B. However, it only looks as if it has reached B'. so the apparent velocity of a receding space ship is slower then the actual speed.

Doppler Distortion

Continuing the discussion of the previous chapter it is interesting to examine the apparent velocity of an object which is moving at a constant speed v along a line which does not pass close to the observer.

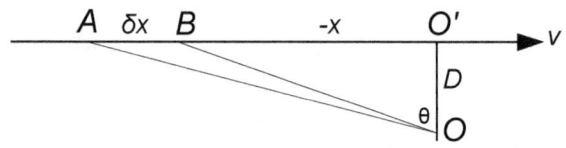

Light from A takes $t_A = \dfrac{\sqrt{(x + \delta x)^2 + D^2}}{c}$ to reach the observer.

Light from B takes $t_B = \dfrac{\sqrt{x^2 + D^2}}{c}$. The time taken by the object to travel from A to B is $t_O = \dfrac{\delta x}{v}$ so the time interval between the receipt of the two pulses emitted at A and B is $t_O - (t_A - t_B)$.

Ignoring second order terms, this works out to be

$\dfrac{x\,\delta x}{c\sqrt{x^2 + D^2}} - \dfrac{\delta x}{v}$ from which it follows that the apparent speed of the object moving from left to right is $v' = v\dfrac{c}{c + v\sin\theta}$ where

$\sin\theta = \dfrac{x}{\sqrt{x^2 + D^2}}$. As the object passes you the $\sin\theta$ term changes from negative to positive and the apparent speed changes from moving faster than v to moving slower.

A graph of this function looks like this:

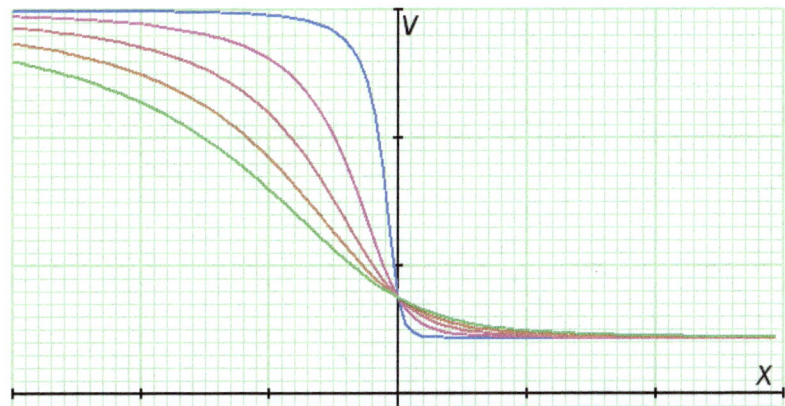

Apparent velocity of a passing space ship travelling at 0.75 c, passing at different distances.

When the space ship is a long way away and approaching the apparent speed is ¾ / (1 − ¾) = 3 times the velocity of light. (The apparent speed of approach will be faster than the speed of light whenever $v > 0.5\ c$. Of course, this does not contradict Special Relativity because nothing is actually moving at this speed.)

At the instant it passes the observer it appears to be going at 0.75 c and slows to an eventual speed of ¾ / (1 + ¾) = 3/7 c. 5 Graphs are shown representing different distances between the observer and the track of the space ship, the dark blue one being the closest.

The graph shows how velocity varies with distance. What we really need is a graph showing how distance varies with time. The algebra is a little comlicated (and can be found in the appendix) but the result is the rather formidable looking function:

$$T = \frac{X}{v} + \frac{1}{c}\left(\sqrt{X^2 + D^2}\right)$$

In units in which c = 1, this function looks like this:

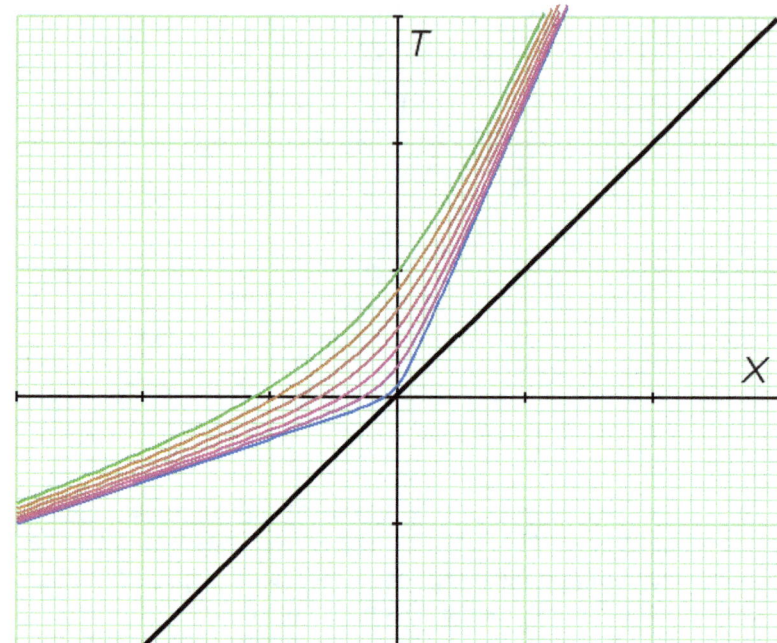

Time/distance graph for an object moving at 0.75 c

Reading this graph from the bottom up, the thick black line shows the actual position of the object the object moving at constant speed from left to right. The coloured lines show where the object appears to be at each instant of time, allowing for the delay caused by the finite speed of light. The object enters on the left, appears to move rapidly to the right (at 3 times the speed of light), slows down as it passes the origin and continues to the right at a slower speed.

The greater the distance *D* the more the object seems to lag behind.

We shall now consider what *shape* an object will appear to be if it is photographed as it moves by. First consider a ruler, perpendicular to the line of velocity which moves past the observer, the ends of the ruler being a distance *D* above and *D* below the centre line. At the instant the centre of the ruler passes the observer, the ends of the ruler appear to lag behind because the light that currently reaches the observer was emitted when the ruler was some distance to the left of the origin.

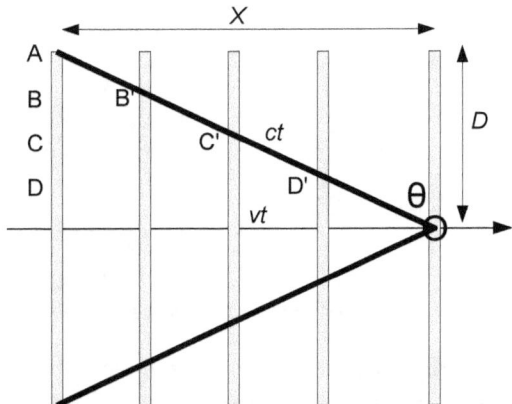

The vertical (blue) lines represent the actual position of the ruler at five instances of time. In the time it takes for light from the end of the ruler A to travel the vertical distance D the ruler has travelled a horizontal distance X so the observer at the origin O sees the end of the ruler at A. By the same token he sees the point B on the ruler at the point B', C at C' and D at D' etc. In other words, he sees down the whole length of the ruler as straight but angled at an angle θ such that $\sin\theta = v/c$. Likewise the bottom half of the ruler also appears angled backwards, the whole ruler therefore looks bent sharply in the middle as shown by the thick black line in the above illustration.

The next question is – what does the ruler look like at times other than $T = 0$. Rearranging the equation on page 75 and solving for X we get:

$$X = \frac{c^2 vT - v\sqrt{c^2 v^2 T^2 + D^2(c^2 - v^2)}}{c^2 - v^2}$$

(When $T = 0$, X must be negative for all non-zero values of D so we need the negative root here.)

The graph of this function looks like this:

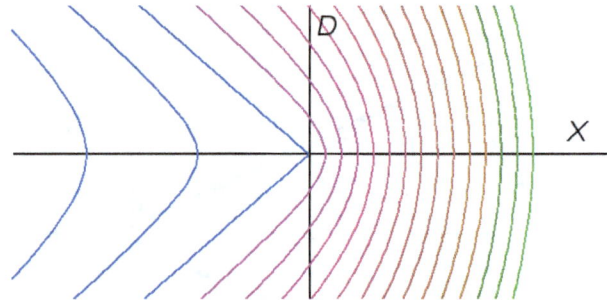

The apparent shape of the ruler at unit intervals

As the ruler approaches the observer at speed, it becomes more and more curved like a boomerang; at the instant it passes the observer it appears to be bent sharply in the middle; then as it recedes more slowly it straightens out again.

Using this diagram as a guide we can picture what the observer sees as a rectangle hurtles past:

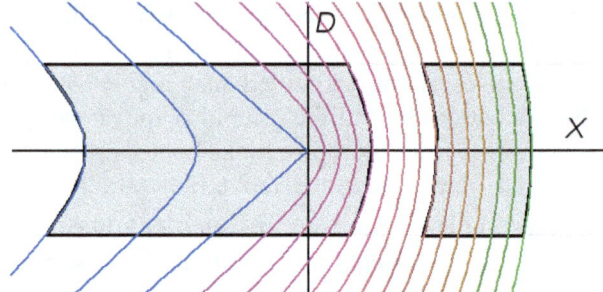

The apparent shape of a moving rectangle

The rectangle is convex at the front and concave at the rear; if shortens as it goes past the origin. The diagram can be rotated around the X axis to give an impression of what a rectangular box or a cylinder would look like. (Once again it is necessary to point out that this is NOT a relativistic effect – it is just a Doppler effect. Factoring in a length contraction does not change anything significant – only the aspect ratio of the rectangle.)

The Doppler distortion effect was first brought to our attention by two papers in the late 1950's independently by Terrell and Penrose. Subsequently many authors have expressed surprise that the apparent distortion of a moving object took so long to be recognised, but the reason it took 50 years for the effect to be noticed is that the effect is, frankly, only of academic interest. It is just an illusion and has no practical or theoretical value. (In fact, I am willing to bet that, since the effect has nothing to do with relativity, it will have been described as a curiosity long before Einstein. Maybe by Doppler himself.)

Here is a picture of a rectangle moving past an observer at a small distance, taken at the instant the centre of the nearer face of the rectangle (looks as if it is) is perpendicular to the observers line of sight.;

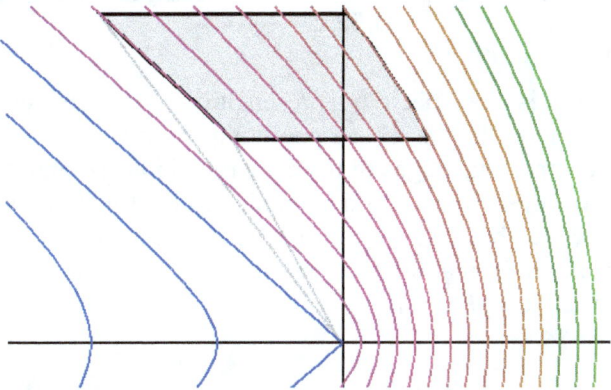

The apparent shape of a moving rectangle

Note that both the nearer and the further faces of the rectangle remain parallel at all times to the axis. (The rectangle could be a flat wagon running on rails) but owing to the finite speed of light the further face of the rectangle looks as if it is lagging behind. In fact, the observer will be able to see the rear face of the rectangle in spite of the fact that it would ordinarily be obscured by the nearer face.

Now, owing to a pretty coincidence, it turns out that, when the length contraction is applied, the apparent angular sizes of the near and rear faces are the same as they would be if the rectangle had been rotated. The effect has therefore come to be known as Terrell Rotation. The above diagram makes it clear, however, that the rectangle is not rotated – it is distorted. Wikipedia has the following illustration of what a cube really would look like as seen from the origin.

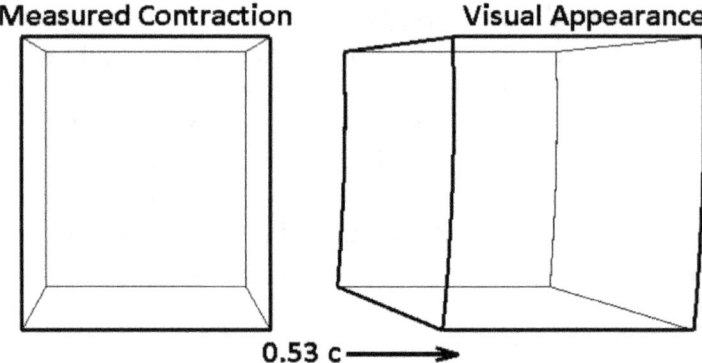

Measured Contraction   **Visual Appearance**

0.53 c ⟶

A Doppler distorted cube

On the left is a perspective representation of what the cube really is –
squashed. On the right is what a photograph would show. Note carefully
that all the horizontal edges of the cube remain horizontal. If the cube
were rotated, these lines would be inclined. It is true that the cube looks
rotated in a way, because we can see the left hand side and our brains
automatically infer that the faces are the same size; but in truth the cube
is not rotated, it is *sheared*.

Part 3: Mass and Energy

Mass and Momentum

When they were children Albert and his twin brother Ludvig had been given two lovely big beach balls and with them they had developed a skilful game. Each boy would throw the ball towards the other and the game was to try to catch the balls as they bounced back. In fact they got so skilful they could do it with two footballs (which are smaller and heavier) as well. One day they tried playing the game with a beach ball and a football. The first time they tried it, the beach ball bounced back off the football – rather fast, actually – and the football just dropped to the ground. Ludvig, who was throwing the football, realised that he was throwing it too hard and that in order to make both balls bounce back at the same speed as they were thrown the lighter beach ball would have to be thrown faster than the heavier football.

Later, when the two boys learned about the conservation of momentum it became clear that because the football had twice the mass as the beach ball the beach ball had to be thrown twice as fast as the football.

Now, as fully qualified space pilots, Albert and Ludvig decided to play their old game but with a new twist. They would each throw a rubber projectile sideways out of their respective space ships as they passed each other at speed and try to recover them when they bounced back.

On the day of the experiment Albert and Ludvig checked to make sure that the two projectiles had exactly the same mass and that the firing mechanism caused the projectiles to be thrown out with exactly the same speed. They synchronised their clocks and measuring instruments and took off leaving their friend Klara back on Earth to take a cine film of the event.

Soon Albert was approaching Earth from the left at a good speed with Ludvig approaching from the right at the same speed. At the appointed moment both space ships ejected their projectiles and, with perfect precision, the balls bounced off and were captured by the nets trailing behind the two ships. All of this was videoed by Klara and later, over a few pints, Albert and Ludvig sat down to watch the video. This is what they saw[21]:

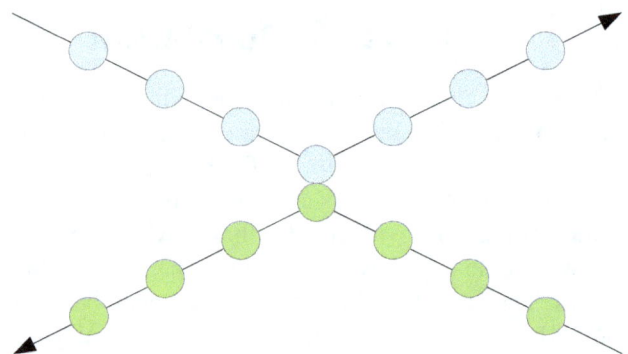

'Fantastic' said Albert; 'Spot on!' said Ludvig as they performed a celebratory High-Five.

'Mind you – that's not what I saw.' said Ludvig. 'From my point of view, my ball went straight up and down while your ball whizzed past at great speed.

'Yes – I saw exactly the same.'

'And another thing.' said Ludvig, 'I saw you release your ball well before I did.'

'But I saw *you* release your ball before me!'

'You're both right.' said Klara. 'Each of you think that your brothers clocks are running slow so each of you will see everything happening on your brother's ship in slow motion.

'So from my point of view, Albert's ball was travelling slower than mine. Is that right?'

'Yes', replied Klara, ' and from his point of view yours was travelling

---

21 Of course, this is not what the video actually recorded. Because of Doppler Dilation discussed in the previous chapter, Klara will have seen the balls approaching at high speed and slowing down as they went past. In accordance with long tradition, the diagram shows what Klara *infers* was happening, not what she *saw*. What is important about the diagram is the symmetry in the behaviour of the two balls. In the subsequent discussion between Ludvig and Albert the word 'see' actually means 'infer'.

slower than his.'

'But – hang on, what about the conservation of momentum?' said Albert.

'Well, obviously it doesn't apply in this situation' replied Ludvig.

'No, it's got to apply' said Klara. 'The law of conservation of momentum is one of the most fundamental laws of Physics. You can't just ditch it like that.'

'But my ball has momentum mv while Albert's ball has momentum kmv which is less, so, like the football and the beach ball, the symmetry is lost and the balls shouldn't bounce symmetrically.'

'Well the situation has got to be symmetrical because I can argue in exactly the same way' Albert pointed out.

'No, I think the answer is a lot more interesting.' said Klara

'How so?' said Albert.

'Well, there is an asymmetry in the situation from your individual points of view. To each of you, your own ball is pretty well stationary (apart from its small sideways velocity) but your brothers ball is whizzing past at high speed.

'You both agree that your brothers ball is moving sideways more slowly then your own; so in order to compensate for that it is obvious that the *mass of the ball must be increased* in the same proportion by virtue of its relativistic forward speed.

'What I am saying is that a ball which when at rest has mass m must have mass m/k when travelling at speed. Now the k's will cancel and the law of conservation of momentum will be upheld.'

'Well that's a cool idea.' said Albert. 'Is there any evidence that mass increases when thing move really fast?'

'I don't know of any.' said Klara.

'I do.' said Ludvig. 'I have noticed a strange thing when Albert is accelerating his spaceship. Even though he has the warp drive on full power the whole time I notice that the acceleration drops off more and more the faster he goes.'

'Yes, but that's just because Special Relativity forbids any object from going faster than light.' said Albert

'It's not just a question of forbidding you to go faster than light. I see now that as you go faster, the mass of your spaceship increases and so the same force produces a smaller and smaller acceleration.'

'Well, that's not how I see it.' replied Albert. 'When I have my rockets on full power I feel a constant acceleration and I just go on getting faster and faster.'

'You mean you can go faster than light?'

'No, of course not.'

'Well if, as you say, you go on accelerating at the same amount, all the time there must come a point, according to you, when you are travelling faster than light.' said Ludvig.

'Hmm, I see your point.'

'You two are talking at cross purposes because you have forgotten to specify what frame of reference you are adopting.' said Klara. 'Ludvig is quite right to say that from his point of view, your speed increases more and more slowly because from his point of view the mass of your spaceship is increasing.

'You, Albert, are also correct is saying that from your point of view your acceleration is constant. But, Ludvig, you are wrong to say that this constant acceleration will result in Albert going faster than light because velocities do not add up linearly.'

'Yes, I had forgotten that.'

'But how does all this tie in with the conservation of energy?' asked Albert.

'Very good question' replied Klara.

Mass, Energy and Momentum

The formula for the relativistic increase in mass for an object of rest mass m moving with a speed v is:

$$M = \frac{m}{k} = \frac{m}{\sqrt{1 - v^2/c^2}}$$

In text books on Relativity this is usually written as γm where $\gamma = \frac{1}{\sqrt{1 - v^2/c^2}}$. Unlike k which is less than 1, γ[22] increases from 1 to infinity as the speed reaches the speed of light.

Now the kinetic energy of an object of mass m moving with a non-relativistic speed v is given by the formula $\frac{1}{2}mv^2$ and you might be forgiven for thinking that the kinetic energy of an object moving with a relativistic speed would therefore be $\frac{1}{2}\gamma mv^2$. Unfortunately this turns out to be incorrect. The correct expression for the kinetic energy of an object of rest mass m moving at a relativistic speed is very simple. It is

$$KE = mc^2(\gamma - 1)$$

At first sight this looks very different but let's see what happens if we unpack it a bit (using a bit of mathematical sleight of hand called the Binomial Theorem):

$$KE = mc^2\left(\frac{1}{\sqrt{1 - v^2/c^2}} - 1\right)$$
$$= mc^2\left(1 + \frac{1}{2}\frac{v^2}{c^2} + \frac{1}{6}\frac{v^4}{c^4} + \dots - 1\right)$$
$$= \frac{1}{2}mv^2 + \frac{1}{6}\frac{mv^4}{c^2} + \dots$$

When v is small only the first term counts and this is $\frac{1}{2}mv^2$ as expected. But we see that there are a whole lot more terms which get more and more significant as v approaches c. This is another reason why Albert cannot reach the speed of light in his space ship. He would need an infinite amount of energy.

Now let's have a closer look at the equation $KE = mc^2(\gamma - 1)$ again. If we rearrange it a bit we get

_____

22 γ is the Greek letter *gamma*.

$$E = \gamma mc^2 = mc^2 + \text{KE}$$

All three terms have the dimensions of energy but what is the significance of each?

γmc^2 is the sum of two energy terms so it makes sense to call this term the '*total relativistic energy*' of the object. This increases with speed and becomes infinite when the speed reaches the speed of light.

mc^2 does not depend on the speed as m is the mass of the object when it is at rest so we might reasonably call this the '*rest-mass energy*' of the object.

We now have a perfectly reasonable proposition: *The total relativistic energy of an object is the sum of its rest-mass energy and its relativistic kinetic energy and is equal to γmc^2.* There is nothing in the mathematics that says that the rest-mass energy actually exists or that we could harness it for purposes good and evil but Einstein speculated that it was so and he was proved spectacularly right on the 16th of July 1945 in the deserts of New Mexico.

Now in the last chapter, Albert and Ludvig discovered that, in order to preserve the law of conservation of momentum, it was necessary to assume that the mass of an object increased as it got faster and that its relativistic momentum p was not just mv but mv/k – or, as it is more usually written, γmv. We have just found out that the *total relativistic energy E* of a moving object is γmc^2. If we write these out in full we get:

$$p = \frac{mv}{\sqrt{1 - v^2/c^2}} \qquad E = \frac{mc^2}{\sqrt{1 - v^2/c^2}}$$

Eliminating v from these two equation is a bit messy but the result is one of the most important equations in Relativity:

$$E^2 = m^2 c^4 + p^2 c^2$$

which gives us the precise relativistic relationship between energy and momentum for any object. (Note that if the object is stationary, $p = 0$ and $E = mc^2$.)

At first sight, this equation looks really strange. For objects moving at non-relativistic speed we have

$$\text{KE} = 1/2\, mv^2 \quad \text{and} \quad momentum = mv$$

which means that, in general $\text{KE} = 1/2\, p^2/m$ which looks nothing like the equation on page 86.

We must, however, remember the E is the *total relativistic energy*, not the kinetic energy. To a first order approximation we can put

$$E = mc^2 + KE$$

in which case

$$E^2 = (mc^2 + KE)^2 = m^2 c^4 + 2mc^2 KE + KE^2$$

Again, to a first order approximation we can ignore the KE^2 term and by putting $KE = 1/2\, p^2/m$ we get the equation of page 86 so it all checks out.

What happens if we apply this equation to light itself?

According to Quantum Theory, light is a stream of particles called photons which have zero rest mass and whose energy is related to their frequency v and wavelength λ by the equation:

$$\boxed{E_{photon} = hv = hc/\lambda}$$

where h is Planck's constant.

Now since photons have zero rest mass we can put $m = 0$ into the energy/momentum equation from which we get:

$$\boxed{E_{photon} = p_{photon}\, c}$$

from which we can deduce that photons must also possess momentum, given by the equation:

$$\boxed{p_{photon} = h/\lambda = hv/c}$$

The energy of a photon of visible light is of the order of 3×10^{-19} J and a powerful torch emitting 3 W of power is producing about 10^{19} (10 million million million) photons every second. The momentum of one of these photons is $E/c = 3 \times 10^{-19}/3 \times 10^8 = 10^{-27}$ kg m s^{-1} and the total momentum generated every second is therefore about 10^{-8} kg m s^{-2} or N. When a fireman holds a hose emitting a jet of water, he has to hold very tightly because the rate of change of momentum of the water is translated into a reaction force which he must oppose. When you hold a torch, you too, much resist the force[23] generated by the issuing photons.

---

23 10^{-8} N is about the weight of a grain of dust!

Albert's Problem

'Do you remember that experiment Ludvig and I did with the two projectiles?' said Albert one day as he was chatting to Klara in the Spaceport lounge.

'You mean when you threw two projectiles out sideways and they bounced off each other?'

'Yes, That's the one. We decided that, in order to preserve the law of conservation of momentum, the mass of a moving object had to increase by a factor of $1/k$' (see page 83)

'Well, I have been doing a bit more reading' continued Albert, 'and I found out that physicists usually use γ rather than k where $\gamma = 1/k$ and that the formulae for the momentum and energy of a mas m moving at a relativistic speed are:

$$Relativistic\ momentum\ =\ \gamma\, mv$$
$$Relativistic\ kinetic\ energy\ =\ mc^2(\gamma\ -\ 1)$$
,

'That's a rather strange formula for relativistic energy' commented Klara, 'but I suppose it makes some sense. When an object is stationary, $\gamma = 1$ so the kinetic energy = 0.'

'I agree – but it seems to check out; at least usually.'

'How do you mean?'

'Well I wondered what some simple interactions would look like from a moving spaceship so I have done some calculations using some simple numerical examples. Shall I show you?'

'Yes, I am intrigued.'

Alberts first example

A mass m is travelling at 60% of the speed of light ($\gamma = 1.25$) and collides with a mass $2m$ travelling in the opposite direction with just the right momentum so that both masses bounce off each other elastically with equal speeds.

In order for the masses to bounce back symmetrically, they must have equal and opposite momentum so, taking the speed of light to be 1 and the masses to be 1 and 2 units respectively and using the relativistic formula for momentum:

$$1.25 \times 0.6 = \frac{2v}{\sqrt{1-v^2}}$$

from which we can calculate that v must be equal to 0.351.

Now the question is, what does this situation look like from the point of view of someone in a space ship? For simplicity, let us suppose that the ship is moving at 60% of the speed of light from left to right. In other words, the smaller mass is stationary and the larger mass is moving at a speed which is the relativistic sum of 0.351 and 0.6. This works out to be 0.786 (using the equation for the relativistic addition of velocities derived on page 42). So in this frame the situation looks like this before the collision:

and like this after the collision:

We can work out the values of v_1 and v_2 by simply adding 0.6 to the bounce back speeds (again using the relativistic formula). This gives us $v_1 = (0.6 + 0.6)/(1 + 0.6\times0.6) = 0.882$ (moving from right to left) and $v_2 = (0.6 - 0.351)/(1 - 0.6\times0.351) = 0.315$ (also moving from right to left because the space ship is moving faster than the mass).

We can now check to see if, in the space ship frame, momentum is conserved.

First we must calculate the γ factors:

$\gamma(0.786) = 1.618$
$\gamma(0.882) = 2.122$
$\gamma(0.315) = 1.054$

Total momentum before the collision = $1.618 \times 2 \times 0.786 = 2.54$
Momentum of small mass after the collision = $2.122 \times 0.882 = 1.872$
Momentum of large mass after the collision = $1.054 \times 2 \times 0.315 = 0.664$

Total momentum after the collision = $1.872 + 0.664 = 2.54$

So everything checks out nicely.'

'What about the kinetic energy? Does that check out too?' asked Klara.

'Well obviously it checks out in the stationary frame because the speeds of the two masses remain unaltered. In the space ship frame the calculations go like this:

Total relativistic KE before the collision = $2 \times (1.618 - 1) = 1.24$
Relativistic KE of small mass after the collision = $2.122 - 1 = 1.122$
Relativistic KE of large mass after the collision = $2 \times (1.054 - 1) = 0.108$

Total relativistic KE after the collision = $1.122 + 0.108 = 1.23$

So, to the accuracy of my calculation at any rate, this checks out too.'

'Well, that's what we should expect isn't it?'

'Yes, but I ran into a problem with my next example.'

'How so?'

Albert's second example

'Consider a mass m travelling at 60% of the speed of light colliding with *and sticking to* a second mass m. This is a simple example of an inelastic collision. We still expect momentum to be conserved so we can calculate the subsequent velocity of the two masses. In fact, the calculation is numerically the same as before and the velocity works out to be 0.351 whose γ factor is 1.068

Since the collision is inelastic, we can expect some KE to be lost and this is, in fact the case.

Total relativistic KE before the collision = $1.25 - 1 = 0.25$
Total relativistic KE after the collision = $2 \times (1.068 - 1) = 0.136$

Total energy lost as heat etc. = $0.25 - 0.136 = 0.114$'

'So what is the problem?' asked Klara.

'Well, in the space ship frame, the fist mass is stationary while the second is approaching at $0.6c$. So the relativistic KE is exactly the same as before – 0.25.

But after the collision, the apparent speed of the double mass is the difference between the two speeds which we have already worked out to be 0.315. The γ factor for this is 1.054 so the relativistic KE is equal to $2 \times (1.054 - 1) = 0.108$. If we add in the heat energy lost we get $0.108 + 0.114 = 0.222$, not 0.25.

'That's not very different.'

'That's not the point. They should be *exactly* the same. There is something seriously amiss here.'

'Well, I am afraid I am not going to be able to help you much – but if your calculations are right it does sound as if the law of conservation of energy is is serious trouble!)

The resolution of Albert's problem

Albert is right. There is something seriously wrong with his calculations. What he has forgotten is that the heat generated by the inelastic collision *also possesses* mass. Relativity requires that mass has energy according to the relation $E = mc^2$. The converse of this is that energy possesses mass equal to E/c^2. The mass of the combined object after the collision is therefore not $2m$ but $2m + E/c^2$ where E is the loss in kinetic energy.

This complicates matters because we cannot work out the final speed of the two masses until we know the energy lost – and we cannot work out the energy lost until we know the final speed. There is nothing for it but to resort to some algebra.

If the velocity of the first mass is v_1 (gamma factor γ_1) and the velocity of the combined mass is v_2 (gamma factor γ_2) then applying the conservation of momentum in the stationary frame, making allowance for the extra mass of the lost energy E (and taking $m = 1$ and $c = 1$) we have:

$$\gamma_1 v_1 = \gamma_2 (2 + E) v_2$$

Now before the collision, the total kinetic energy is simply ($\gamma_1 - 1$).

After the collision we have the relativistic KE of the combined masses $(2 + E)(\gamma_2 - 1)$ plus the lost energy E so

$$\gamma_1 - 1 = (2 + E)(\gamma_2 - 1) + E$$
$$\gamma_1 + 1 = \gamma_2(2 + E)$$

Eliminating E from these equations we get:

$$\gamma_1 v_1 = \gamma_2 \frac{(\gamma_1 + 1)}{\gamma_2} v2 = (\gamma_1 + 1) v2$$

$$v_2 = \frac{\gamma_1 v_1}{\gamma_1 + 1}$$

Since in our example, $v_1 = 0.6$ and $\gamma_1 = 1.25$, it follows that $v_2 = 0.333$ ($\gamma_2 = 1.061$), slightly slower than the figure of 0.351 which Albert calculated because of the extra mass of the heat energy.

We can also calculate this energy E which is

$$E = \frac{\gamma_1 v_1}{\gamma_2 v_2} - 2 = \frac{1.25 \times 0.6}{0.33 \times 1.061} - 2 = 0.121$$

Now for the accounting in the space ship frame.

The total relativistic KE before the collision is as Albert calculated it — that is 0.25

After the collision, the KE of the combined mass is

$$(2 + E)(\gamma_2 - 1) = 2.121 \times 0.061 = 0.129$$

Adding in the lost energy $0.129 + 0.121 = 0.25$!!

It seems truly miraculous that, including both the momentum and the mass of the 'lost' energy, the law of conservation of energy is upheld in both frames.

Of course, just proving that the law is true in one simple instance does not prove the general law but it is indeed true in all frames and in all circumstances.

But it all goes to show just how subtle relativity is.

Albert's example illustrates a very important point: *anything which has energy has extra mass too*. A hot cup of tea is more massive than a cold cup of tea; A wound up wrist watch is heavier than when it is run down; A pair of cylinders containing $2N$ atoms of hydrogen and N atoms of oxygen are more difficult to move about than the same cylinders filled with N molecules of water; the combined mass of a cuckoo clock plus the Earth is greater when the weight has been pulled up than when it has reached the bottom[24] etc. etc.

In all these cases, the more energetic system has more mass by any definition you can think of. They *weigh* more in a gravitational field and they possess more *inertia* when a force is applied. Energy really does have mass (though, of course, in all the cases cited the magnitude of the mass is tiny.)

Even photons have mass[25].

A box with perfectly reflecting walls containing a few photons buzzing around will weigh more than one without[26].

---

24 Note, it is not the *clock* which is heavier; it is the Earth/clock *system* which is more massive.

25 Of course, the *rest-mass* of a photon is zero but it has a relativistic mass of $h\nu/c^2$.

26 Just such a box was the subject of a famous debate between Einstein and Bohr in 1930

The Doppler Shift and the Photon

In a previous chapter we derived an expression for the Doppler shift of light assuming that light was a wave (see page 66). But if light is not a wave at all but a stream of photons, how can they be affected by the relative motion between the source and the observer? After all, if you are travelling towards a source of photons you will still see them hit you at the speed of light so what is the difference?

The answer is obvious. The photons will still appear to you to be travelling at the same speed *but they will have more energy and more momentum.*

This is a very pretty idea – but does it actually check out numerically? Lets start with a more down-to-earth example.

A terrorist is travelling towards you on a flatbed truck at a relativistic speed v firing bullets from a machine gun which is known to have a muzzle velocity of u. The question is – what is the energy and momentum of the bullets when they hit you?

Well, assuming that you are still in a position to care, the calculations go like this: first it will be convenient to work out a formula for the γ factor appropriate to an object moving with a speed equal to the relativistic sum of u and v.

$$\gamma_{u+v} = \frac{1}{\sqrt{1 - \left(\dfrac{u + v}{1 + uv/c^2}\right)^2 / c^2}}$$

This formidable expression does, however, simplify quite nicely and reduces to:

$$\boxed{\gamma_{u+v} = \gamma_u \gamma_v (1 + uv/c^2)}$$

(The proof of this formula will be found in the Appendix.)

Armed with this useful formula we can now easily write down the momentum and energy of the bullet:

$$\text{momentum } p = \gamma_{u+v}\, m\, \frac{u + v}{1 + uv/c^2}$$
$$= \gamma_u \gamma_v m(u + v)$$
$$\text{total relativistic energy } E = \gamma_{u+v}\, mc^2$$
$$= \gamma_u \gamma_v m c^2 (1 + uv/c^2)$$
$$= \gamma_u \gamma_v m(c^2 + uv)$$

In themselves, these formulae are, perhaps, not particularly important. What is more interest is the *ratio* of the momenta and energies in the two frames.

According to the terrorist:

$$\text{original momentum } p_0 = \gamma_u m u$$
$$\text{original energy } E_0 = \gamma_u mc^2$$

so the respective ratios are:

$$\frac{p}{p_0} = \gamma_v \frac{u+v}{u}$$
$$\frac{E}{E_0} = \gamma_v(1 + uv/c^2)$$

Just to fix these ideas more firmly, if the truck was moving towards you at a speed of $0.6c$ ($k_v = 0.8$) and the muzzle velocity was also $0.6c$ then the momentum of the bullet would be increased by a factor equal to $0.8 \times 1.2/0.6 = 1.6$ and the energy would be increased by a factor $0.8 \times (1 + 0.36) = 1.09$.

You could reasonably argue that all this is pretty academic, considering the circumstances, but what is important about these formulae is not what they contain, but what they are missing.

The formulae do not mention the mass of the bullet!

The implication of this is that the formulae are just as valid for bullets of zero rest mass (i.e. photons) as they are to real bullets. All we have to do to get the equivalent formulae for photons is let u tend towards c. This gives us for momentum:

$$\frac{p}{p0} = \gamma_v \frac{c + v}{c} = \frac{c + v}{c\sqrt{1 - v^2/c^2}}$$

$$= \frac{c + v}{\sqrt{(c - v)(c + v)}}$$

$$= \sqrt{\frac{c + v}{c - v}}$$

and for the energy

$$\frac{E}{E0} = \gamma_v(1 + v/c) = \frac{c + v}{c\sqrt{1 - v^2/c^2}}$$

$$= \frac{c + v}{\sqrt{(c - v)(c + v)}}$$

$$= \sqrt{\frac{c + v}{c - v}}$$

which is exactly the same!

On reflection, this is just as it should be because, for a photon, $E = pc$ so energy and momentum are proportional.

In addition, we have arrived at exactly the same formula for the Doppler shift of a photon that the guard and the station master worked out assuming that light was a wave!

Indeed, it may be said with equal verity that the Doppler shift in light is due either to a combination of the Moving Source effect and Time Dilation assuming that light is a wave *or* to the relativistic addition of velocities assuming that light is a stream of particles.

In this instance, Relativity and Quantum Theory are seen to be in perfect agreement.

Part 4: Gravity

The Experiment in the lift

Arthur and Betty were visiting Shanghai and Arthur suggested they go to the observation deck at the top of the Shanghai Tower.

'Great idea.' said Betty. 'Wait a mo while I get some things, though.'

Soon they were waiting in the lobby for the lift to come down. Betty rummaged in her rucksack and pulled out a bathroom scales and a stop watch.

'What on Earthy do you want those for?' queried Arthur.

'I want to measure the height of the tower,' she answered.

'How are you going to do that? Wait – I have it. You are going to throw the scales off the deck at the top and time how long it takes to reach the ground! Is that it?'

'No, silly.'

'Well how then?'

'Wait and see.'

When the lift arrived, Betty placed the scales on the floor and asked her brother to stand on them. Then she crouched down to look at the pointer. It read 80 kg.

'Now press the button for the top floor, Arthur' she said.

As the lift ascended, Betty scribbled some figures down in her notebook.

When they got to the top and had had a look at the fabulous view of the city, Arthur said 'What was all that about in the lift on the way up?'

'Well, what did you notice when the lift started accelerating upwards?'

'It felt as if I got a bit heavier.'

'Yes, the scales read an average of about 96 kg for 10s.'

'What can you deduce from that?'

'Well Newton's law says that force = mass × acceleration. We know that you normally weigh 80kg. The extra 16 kg force was accelerating you upwards.

'As 16 kg force is 160 N the acceleration of the lift must have been $160/80 = 2$ ms^{-2}.

'Since the acceleration lasted for 10 s, I can deduce that the lift achieved a speed of 20 ms^{-1} which is about 45 mph.'

'Wow – that's pretty impressive. The speed, I mean; not your calculations,' said Arthur.

'Thanks a lot.'

'Okay – I didn't mean it. How long did we travel at that speed?'

'15 seconds.'

'So we must have travelled 300 m.'

'Correct – and during the accelerated phase we travelled about 100 m.'

'What did the scales read while we were travelling upwards at constant speed?' asked Arthur. 'I suppose it was a bit more than 80 kg because it would need a bit of force just to keep me moving.'

'No, you are completely wrong about that. It is true that a car, for instance, needs a constant force just to keep it moving but that is because there is always friction. But for you in the lift there is absolutely no friction because the air all around you is moving at the same speed. No, the scales read exactly 80 kg.'

'Yes, I see that now.' Arthur conceded.

'What did you experience when we were approaching the top?'

'My stomach gave quite a lurch!'

'That is because you were experiencing negative acceleration, Negative g is always more uncomfortable than positive g. That is why the designers of the lift slowed the lift down in 20 seconds not 10.'

'What was the reading on the scales?'

'The scales read 72 kg which means that the deceleration was 1 ms^{-2} and during that time we travelled another 200 m making the total distance, and the height of the tower $100 + 300 + 200 = 600$ m.'

'Well, I have to admit, I am impressed after all. And to think that you were able to calculate all of this without looking outside the lift!'

'That's a really important point. You know that, when you are travelling in a train, it is impossible to determine how fast you are going simply by doing experiments inside the train – indeed, that is the fundamental principle of Special Relativity. But when it comes to frames of reference which speed up, slow down, or turn a corner, it *is* possible to measure the acceleration.'

Later that day, Arthur thought hard about what his sister had said and he had an idea.

'I don't think you are right to say that you can always measure acceleration by doing experiments inside a closed box.'

'Why not?' said Betty.

'Because I can think of a way of getting the same results without moving the box at all!'

'How could you do that?'

'By switching on some extra gravity!' said Arthur triumphantly.

'How?'

'Well, in order to make you think you were accelerating, I could just quickly move a large asteroid under the box for a while.'

'Sounds rather unlikely to me. But actually you are perfectly right. Inside the box acceleration and extra gravity are completely indistinguishable. Astronauts in some future space ship could simulate Earth's gravity by simply accelerating at 10 ms^{-2} and they wouldn't be able to tell the difference at all.' said Betty.

'Maybe gravity doesn't exist and we only feel a downward force on our bodies because the Earth is actually expanding at 10 ms^{-2}!'

'Don't be ridiculous!' said Betty.

'I am not being ridiculous. Can you prove that it isn't?'

'Well, no, actually; I suppose I can't.'

'There you are then' said Arthur triumphantly.

Newton's Equivalence Principle

Galileo was the first person to realise that, if you could only remove all resistance, bodies of different mass would all fall with the same acceleration and Newton was the first person to explain why. The reason was, he argued, that the weight of a body (i.e. the force of gravity on it) was *proportional* to the mass but the acceleration of a body (acted upon by a constant force) was *inversely proportional* to the mass. The acceleration of a body just accelerated by its own weight would therefore be independent of the mass and in the case of the acceleration due to gravity at the surface of the Earth it would be 9.8 ms^{-2}.

There is, however, a hidden assumption in this argument, as Newton himself realised. It assumes that gravitational mass and inertial mass are identical. This may seem obvious but there is no logical necessity for it to be so. One could imagine, for example, that the force of gravity could act differently on different elements – for example iron and aluminium – and that the force of gravity on 1 kg of aluminium could be different than the force of gravity on 1 kg of iron[27].

'That's silly!' said Arthur when Betty had explained this to him. 'If I want to measure out 1 kg of aluminium rivets I get a pair of kitchen scales, put a a 1kg iron weight on one side and then pile up aluminium rivets on the other until the scales balance. Obviously, then, the force of gravity would be equal on both objects.'

'True, but if gravity really did act more on, say, aluminium than it does on iron, then you wouldn't need to put 1 kg of rivets in the pan to make the scales balance.'

'OK then, suppose I use a bathroom weighing machine instead of scales. Are you saying that when I put my rivets on the weighing machine they will weigh less than 1 kg?'

'No I am not saying that. The weighing machine, like the scales actually measures the force of gravity on an object, not its actual *mass*.'

'So how can I compare the mass of two objects without using

---

27 When I talk about 1 kg of aluminium and 1 kg of iron, I am referring to the *inertial* mass of the substance. Inertial mass is more obviously connected to the 'amount of matter' present because it is obvious that inertial mass does not change if, for example, to take it to a place where the strength of gravity is different or even zero.

gravity?'

'You can use artificial gravity. You remember when we were discussing what it felt like to be accelerated in the lift? What do you think would happen if we had put a pair of scales on the floor of the lift with a 1 kg mass of iron in one pan and a pile of rivets in the other so that the scales balanced when the lift was stationary?'

'I don't see why the scales should do anything other than remain balanced as the lift accelerated upwards.'

'That is what happens usually,' conceded Betty, 'but if gravity acted more strongly on aluminium than iron, I contend that there would be *less* than a kilogram of rivets in the pan.'

'So what would happen?'

'Well the upward force needed to accelerate the aluminium would be less then the upward force needed to accelerate the iron and the scales would tip towards the iron.'

'That's weird.'

'It may be, but it is not illogical. Consider this. How would you measure out a kilogram of rivets if you were in a space ship, far away from any gravitating planets or stars?'

'What equipment have I got?' asked Arthur

' Anything you like. A pair of scales; some kg weights, a bathroom weighing machine – and, of course, a space ship equipped with rocket engines.'

'Well, I know it is no use using the scales and the weighing machine in zero g because in zero g, the rivets don't weigh anything; so, I guess I will have to fire up the engines. Then with artificial gravity produced by the acceleration, I could use the bathroom weighing machine as normal.'

'Yes, that's right. Under artificial gravity the weighing machine would register a reading, but it would only register the right reading if the rocket engines produced an acceleration of exactly 1G. You can easily get round this problem, though, by putting the kg mass on the machine first, noting the reading, then piling on the rivets until the reading was the same. The it wouldn't matter what the acceleration of the rocket was.'

'That's cool. Come to think of it, it would be even easier to use the kitchen scales. If I understand you correctly, if the pile of rivets in the pan exactly balances a kilogram mass in the other pan, then the masses will be exactly equal, no matter what the acceleration of the rocket is.'

'Exactly so.'

'And when we got back to Earth and weighed the rivets using Earth's gravity, the rivets would *weigh* more than the kg mass; and if I dropped the rivets they would fall faster than the iron – is that right?'

'Right again. But, of course, the remarkable thing is that this doesn't happen. Experiments have been performed which show that inertial mass and gravitational mass are identical to one part in several billion – but nobody knows why.'

'Well, it seems obvious enough to me.'

'How so?' said Betty.

'Because inertial mass and gravitational mass are the same thing!'

'Well, that is perhaps, a rather naïve way of putting it but you are essentially right. If you can't *explain* why A equals B you can always put it on a pedestal and proclaim that you have discovered a new fundamental principle A = B! Which is basically what Einstein did when he claimed that it was impossible to tell the difference between an accelerated frame of reference and a uniform gravitational field.'

'That sounds very much like the principle that led him to Special Relativity, namely, that it is impossible to tell the difference between a stationary frame of reference and a moving one.'

'Yes, but if you thought that Special Relativity was bizarre, his General Theory of Relativity was even stranger! Did you know, for example that gravity bends light?'

The Bending of Light

There is another experiment that Betty could have performed to measure the acceleration of the lift. She could have set up a laser pointer to shine a beam of light across the lift and noted exactly where it fell. Then she could note where the laser beam fell when then the lift was a) accelerating, b) travelling at constant speed and c) decelerating.

Let us consider the constant speed case first.

At first sight, you might think that the laser beam would be deflected downwards because in the time it takes for the light to cross the lift t, the lift has moved upwards by a distance vt and so the beam should hit the opposite wall this distance lower. This conclusion is false[28]. If it were the case, then Betty would be able to make an absolute measurement of the speed of the lift – something which is explicitly denied by the fundamental Principle of Relativity – but the reason why this conclusion is false needs some explaining.

The error most people make is to suppose that having set up the laser pointer so that it directs a ray of light at right angles to the wall, the beam remains at right angles to the wall even when the lift is moving upwards. This is not the case.

When Betty sets up the laser pointer to direct its beam parallel to the floor when the lift is stationary, she is in fact lining up three points; 1) the source of the light, 2) the collimating device and 3) the mark on the wall. Essentially the arrangement is as follows:

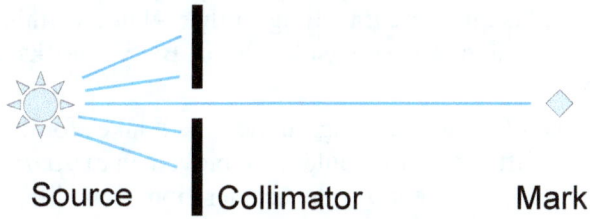

Source　　　Collimator　　　Mark

Light from the source is emitted in all directions but the collimating device (which in a solid state laser is the actual crystal itself which emits the light) only permits the horizontal beam to pass and it is this ray

---

28　The number of websites which claim this to be true is too large to enumerate.

which hits the mark on the wall.

Now consider what happens if the whole apparatus is moving upwards at a constant speed.

In the time that it takes for the light to cross from the source to the collimator, the collimator itself has moved upwards. The horizontal ray will be blocked and it will be an upward moving ray which is allowed to pass through – and, of course, this is precisely the ray which will go on to hit the mark on the opposite wall!

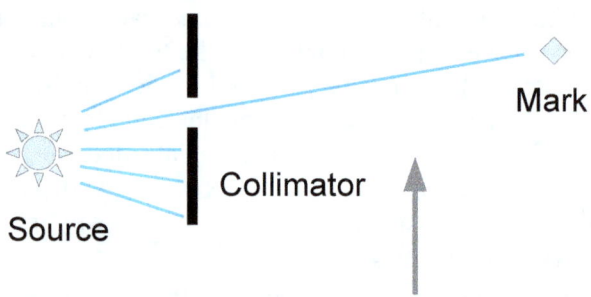

It is true that, if a light beam were directed through a hole in the lift from *outside*, the light would hit the opposite wall a distance *vt* below the horizontal, but Betty's laser pointer is *inside* the lift and partaking of the same velocity and it will always generate a ray which hits the opposite wall in the same place regardless of the speed which which the lift is moving (provided that the speed is constant, of course). To an observer outside the lift, it will appear that the light is travelling along a diagonal line and is therefore travelling further – but then this is exactly the reason why, to an observer outside the lift, Betty's clocks seem to be running slow[29].

But when the lift is accelerating, in the time it takes for the light to travel across the lift *t*, the lift would have moved an *extra distance* upwards equal to $\frac{1}{2}at^2$ where *a* is the acceleration of the lift.

From an outsider's point of view the light beam travels in a straight line but to those inside the lift, it appears to bend downwards and by measuring the degree of bending[30] (!), Betty could easily calculate the

---

29 For a simple proof of the Time Dilation formula based on this insight, see the appendix.

30 If the lift was 3 m wide, the time taken for the light to cross the lift would

acceleration of the lift.

The example of the lift is complicated by the fact that both real and artificial gravity play a role. To simplify things further we must imagine Betty in a space ship measuring the acceleration of the ship in exactly the same way by measuring the bending of a light beam compared to the position of the beam when the rocket engines are switched off. Let's suppose that the ship is accelerating at 10 m s^{-2} – in other words at 1G, a fact which Betty can confirm by standing on a bathroom weighing machine and noting that it reads exactly the same as it did back on Earth. The question now arises – when Betty lands back on Earth, will the light beam bend or not?

Virtually all nineteenth century physicists would say no. Light was then believed to be a wave and this was emphatically confirmed by James Clerk Maxwell who showed that it was electromagnetic in origin. Nothing in his theory gave any grounds for supposing that light would be affected by gravity.

Newton, however, would have disagreed. For him, light was a stream of particles which, like everything else, were subject to his universal force. He even used this idea to show how light would bend when entering a dense medium such as glass.

Einstein sided with Newton, but not because he thought that light was a stream of gravitating particles, but for a very different reason. It seemed to him that just as it had proved impossible to detect absolute motion through space, it ought to be impossible to prove that you were accelerating. This is Einstein's Equivalence Principle:

> *It is impossible by carrying out experiments in a small closed laboratory to distinguish between uniform acceleration and a uniform gravitational field.*

A number of important facts can be deduced from this assumption. The first is that gravity bends light in exactly the same way that it bends the paths of any ordinary projectile and that when Betty returns to Earth she will find the laser beam deflected by exactly the same amount that it was deflected when the space ship was accelerating at 1G.

---

be 10^{-8} s and the spot would therefore deflect something like 10^{-16} m – which is smaller than the diameter of a proton!

The second is that gravitational mass and inertial mass must be *exactly* the same. If Arthur's pile of rivets and his 1 kg mass 'fall' to the back of the spaceship with exactly the same acceleration (which they must do because they are not actually accelerating – it is the space ship which is accelerating in the opposite direction) then they must fall to the ground in Earth's gravitational field with the same acceleration too. This means that inertial mass and gravitational mass *must* be identical.

Thirdly, it follows that in any frame which is in 'free fall' – that is to say, any frame which is accelerating under the influence of a local gravitational field such as a lift whose cable has broken or a satellite in orbit round Earth, the laws of physics will appear to be identical to the normal laws of physics experienced by an observer far from any gravitational fields.

Black Holes

'I like the idea that gravity bends light,' said Arthur. 'Could you put a light beam in orbit, say, round the Sun?'

'Yes, you could in principle,' said Betty, 'but because light travels so fast it would have to be very close to the Sun.'

'How close?'

'Well the standard expression for the speed of a satellite orbiting a star of mass M at a radius R is $v^2 = GM/R$ where G is Newton's Gravitation Constant. So $R = GM/v^2$.'

'Do you know the value of G?'

'Not off hand; and I don't know the value of M_{sun} either but I do know that Earth orbits the sun in 1 year and that light takes 8 minutes to reach us from the sun.'

'How does that help?'

'Well, it means that the circumference of the orbit is about 50 light-minutes or $1/10,000^{th}$ of a light year and that the speed of the Earth round the Sun is about $1/10,000^{th}$ of the speed of light.

'Since the radius of the orbit is inversely proportional to the speed squared, it follows that the radius of the orbit we are seeking is 100 million times smaller than the orbit of the Earth – which is about a mile.'

'That's no good. I think the Sun is a bit bigger than that,' said Arthur.

'Yes, but black holes are thought to be objects in which a mass is compressed to essentially a point, so light could orbit round a black hole whose mass was equal to the mass of the Sun at a radius of about 1 mile.'

'That's pretty cool!'

The radius is called the Schwartzschild radius and you can easily calculate it for any mass using the formula $R_{schwartzschild} = 2GM/c^2$.'

'Hang on – did you say *two* GM/c^2?'

'Yes, I did.'

'But I thought you said the formula for the radius was just GM/c^2.'

'Well, I was hoping you wouldn't notice that.'

'So?'

'The truth is, I don't really understand it,' confessed Betty. 'But it turns out that in a case where an object is being deflected round a star, the bending is exactly twice the bending you get when you are just in a uniform gravitational field. So with twice the bending, you can orbit the black hole at double the distance. The Schwartzschild radius of the Sun is more like 2 miles, not 1.'

'But the bending of light in a uniform field is as predicted by Newton?'

'Yes – that is guaranteed by Einstein's Principle of Equivalence. There is no doubt about that.'

'I would still be interested to know where this extra factor of 2 comes from, though.' said Arthur.

'So would I,' agreed Betty.[31]

---

31 In fact the situation is even worse than Betty thinks. In the first case, when you get close to a black hole, space is so distorted it becomes meaningless to talk about the 'radius' R of an orbit. It is, however, possible to talk meaningfully about the *circumference C* of an orbit and we can usefully define the 'radius' as being equal to $C/2\pi$ – but we must abandon any thought of the 'radius' as being in any way connected to 'the distance between the orbit and the centre of the black hole'. Secondly, it turns out that the 'radius' at which light can orbit a black hole is actually $3GM/c^2$ i.e. 1.5 times the Schwartzchild radius.

Albert's Slow Clock

There is a third experiment which Betty could have done to measure the acceleration of the lift (or, more realistically, the acceleration of a space ship). Special Relativity predicts that, when the space ship is accelerating, a clock at the back of the ship will run more slowly than a clock at the front so all Betty has to do is to prepare two very accurate synchronised clocks at the back of the ship; then she moves one clock to the front of the ship for a few minutes; then she moves the second clock up to the front. The two clocks will now differ in the time they read and it is a simple matter to calculate the acceleration of the space ship.

'But why on earth should a clock at the back of the ship run more slowly than a clock at the front?' exclaimed Arthur.

'It is all due to the Doppler effect.' said Betty. 'Suppose that the clock at the back of the ship sends timing signals to the clock at the front of the ship every second. Let us also suppose that the ship is l metres long and that it is moving at a constant speed of v ms^{-1} where v is much smaller than the speed of light c.

'Consider the situation from the point of view of an inertial observer (i.e. one who is not being accelerated) outside the ship. The external observer sees each light pulse travel the length of the ship *plus the extra distance travelled by the ship in this time*. It is easy to see that the time taken will be $l/(c - v)$. What this means is that the clock at the front of the ship will receive the timing signals at 1 second intervals but delayed by an amount $l/(c - v)$, and if the ship is travelling at constant speed the delay will remain constant too.

'Now if the space ship is accelerating ...'

'I see what you are getting at,' said Arthur, ' If the ship is accelerating all the time, the delay keeps getting longer and longer; so, the times of arrival of the light pulses as recorded by the clock at the front get progressively later and later.'

'That's right. Each second the speed increases by the acceleration a so the time delay increases by

$$\frac{l}{c - (v + a)} - \frac{l}{c - v}$$

'This looks a bit complicated but if we remember that v is much smaller than c it boils down to just al/c^2. What this means is that every second as measured by the clock at the back turns into $1 + la/c^2$ seconds at the front. To the clock at the front it looks as if the clock at the back is running slow.'

Betty's simple argument led to the prediction that the clock at the back of the space ship would run more slowly by a factor of $1 + la/c^2$. But she assumed that at all times the space ship was moving much more slowly than the speed of light. If the analysis is carried out more carefully it turns out that the correct formula for the time dilation factor is $\dfrac{1}{\sqrt{1 - 2la/c^2}}$. This works out to be exactly the same as Betty's formula when $2la$ is much smaller than c^2 and second order terms can be ignored.

Of course, it is no accident that this formula looks very similar to the standard formula for the relativistic time dilation factor which is $\dfrac{1}{\sqrt{1 - v^2/c^2}}$ just with the v^2 replaced by $2la$.

'I see,' said Arthur. 'So if clocks at the back of a space ship go slow I suppose rulers will also be contracted by the same factor.'

'No, that's not correct,' said Betty. 'Suppose Albert, at the back of the accelerating space ship is measuring the speed of light by measuring the time it takes for a beam of light to travel along a standard metre rule placed at right angles to the direction of motion of the ship. Ludvig, at the front of the ship can just look down the space ship and *see* that Albert's metre ruler is the same length as it always was. (If Albert and Ludvig were on a railway train, Albert would always measure the width between the rails correctly however fast or with whatever acceleration the train was going.)'

'But surely, ' objected Arthur, ' that means that, according to Ludvig, Albert will not get the right answer for the speed of light – a result which contradicts the basic principle of Special Relativity, that all observers must gate the same answer for the speed of light, however they are moving.'

'No. Albert will, of course, get the right answer when he makes his measurements and does the calculation. It is Ludvig who infers that, since, to him, his twin brother's clock is running slowly, light actually does travel more slowly at the back of the spaceship than at the front and that Albert's calculation is in error.'

'I don't get that. How can light travel slow for Albert but at the right speed for Ludvig?'

'For the same reason that time travels more slowly for Albert than it does for Ludvig. But don't forget that Albert will insist that the speed of light is exactly what it should be and that, to an inertial observer outside the ship, light will appear to travel at the same speed at both ends of the ship. No – the only reason Ludvig thinks that Albert's clocks are going slow is because the pulses produced by Albert's clock are less frequent than his own, because every second the light has to travel a bit further to get from the back of the ship to the front.

'So let me get this right,' said Arthur. 'Relativistic length contraction and time dilation are real effects because they can make pennies bend, ropes break and twins to age differently. But clocks running slow at the back of a space ship, that's just a Doppler-shift illusion and isn't real. Is that right?'

'No, that's not right,' said Betty. 'If some time later Albert brings his clock up to the front of the space ship, his clock will not read the same as his brother's. It will have genuinely lost time and, in exactly the same way that he returned from his journey to Alpha Centauri younger than his brother, he will be that bit younger in this case too.

'You've got to be kidding!'

'No I'm not. There is no way that Albert's clock can 'catch up' with Ludvig's when he brings it up to the front of the space ship. Any time lost while it was at the back during the accelerating phase is lost permanently and Albert will himself be a few seconds younger than his twin brother.'

'Really? I find that difficult to believe.'

'It's true though,' said Betty.

'I seem to remember' said Arthur ' that the twins only differed in age when they got back together because there was an asymmetry in the situation so I suppose that, instead of Albert joining his brother at the front, if Ludvig joins his brother at the back, they will be the same age, won't they?'

'I am afraid not. It doesn't matter who joins who, the damage has already been done and Albert is going to be younger either way.'

'That's weird.'

'But there is even more significance to this thought experiment,' said Betty.

'What is that?'

'You remember that Einstein's Principle of Equivalence laid down the rule that it is impossible to tell the difference between an accelerated system and a gravitational field?'

'I do.'

'Well, that means that if clocks at the back of an accelerated space ship run slow – so do clocks at the bottom of a well!'

'Are you telling me that if Albert went and lived at the bottom of a well like a hermit for a few years, he would emerge younger than his twin brother?'

'I mean exactly that!' replied Betty.

'Is that because gravity makes clocks run slow?'

'No, not quite. Throughout this discussion we have been assuming that both clocks experience the same acceleration in the space ship and the clock at the bottom of a well experiences the same strength of gravity as the clock at the top. So it is not the presence of gravity which in itself causes the clock at the bottom of the well to run slow.'

'What is it then?' asked Arthur.

'You know that gravity bends light so, as the light pulses from the clock at the bottom of the well climb up ...'

'I've got it! They slow down like projectiles from a gun!'

'No. They can't slow down. Light always travels at the same speed regardless of who measures it.'

'So how does gravity affect the light then?'

'It reduces its *energy* and hence also its *frequency*.'

'Of course! It is Doppler shifted! Just like the danger signal.'

'That's right. When light climbs up through a gravitational field like the field surrounding a star, it loses energy and becomes red-shifted.'

'Is that why the distant galaxies show a red shift?'

'No. That happens for an entirely different reason. In fact the gravitational red shift from an ordinary star is almost too small to be measured; but, of course, things are different when it comes to black holes. Round a black hole the gravitational field is so intense that any photon which tried to leave its surface would be red-shifted out of existence.'

'You mean, it wouldn't have enough energy to escape?'

'Exactly that.'

'Is it possible to measure the slowing down of clocks at the bottom of a well?' asked Arthur.

'Yes it is possible but the effect is very small.'

'How small?'

'Well to get some idea we can use my formula but just put the depth of the well for l and the acceleration due to gravity for a. For a well 10 m deep this gives us a dilation factor of approximately $1 + 10 \times 10 / 300,000,000^2$. So a clock at the bottom of this well would lose a second in about 29 million years!'

'Well I don't need to worry about that then!'

'But your smartphone does when it calculates its position by receiving incredibly accurate timing signals from the GPS satellites orbiting 20,000 km overhead. Unfortunately, however, we cannot use my simple formula to calculate the size of the effect because gravity is less at that height. I am sure it can be done somehow though.'

Gravitational Time Dilation

Betty is right. But we need a new idea here – the idea of gravitational *potential*. When you lift up a mass m from A to B through a height h in a uniform gravitational field g, you give it gravitational potential energy mgh. Double the mass and you double the energy. The quantity gh is called the *gravitational potential difference* between the two points and is equal to the gravitational potential energy *per unit mass*. It is usually given the symbol $\Delta\varphi$. φ stands for 'potential' and Δ stands for 'difference'.

If the gravitational field is not uniform, then we have to add up all the little bits of potential difference between A and B. This process is called integration so in general we have:

$$\Delta\varphi = \int_A^B g\,dh$$

The result of this is that the difference in gravitational potential between the surface of a planet of radius R and a point at a height h above it is equal to $\Delta\varphi = \dfrac{Rh}{R+h}g_s$ where g_s is the acceleration due to gravity at the surface.[32]

In the case of a GPS satellite, $h = 20,000$ km, $R = 6400$ km and $g_s = 10$ ms^{-2}. This works out to about 50 million joules per kilogram. (Which gives you some idea of the amount of energy needed to put a satellite into orbit.)

The correct formula for the Gravitational Time Dilation factor[33] is

$$\frac{1}{\sqrt{1-2\Delta\varphi/c^2}}$$

which, as we have seen, approximates to $1 + \Delta\varphi/c^2$ so, in a year, say, (31 million seconds) the clock in the satellite will gain on the clock on the ground by $31,000,000 \times 50,000,000 / 300,000,000^2$ seconds or 17 milliseconds. Not a lot!

It is also worth pointing out that, as the satellite is moving at a speed

---

32 For a proof see the Appendix.
33 A proof of this formula will be given in the chapter on the Rotating Space Station on page 124.

of about 4 km s^{-1}, there will also be a time dilation effect due to its speed. As it happens, this turns out to be a bit smaller – about 6 milliseconds per year, so the clock in the satellite will gain slightly more more than it loses.

As Betty said, the presence of a gravitational field does not of itself makes clocks run slowly. Neither does motion. Both these effects only exist *relative to another observer* who is either in a different place in the gravitational field or in relative motion. When Ludvig receives messages from his brother on his journey to Alpha Centauri, he *deduces* that Albert's clocks are running slow but he cannot confirm that until Albert returns home and the discrepancy is revealed. In the same way, Ludvig can *deduce* that the clock at the bottom of the well is running slow but again this is only confirmed when the clock is raised to the surface again.

There is, however, one crucial difference between the effects of gravity and the effects of motion. Both cause clocks to run slow – but *only motion causes lengths to contract*. This enables the guard and the station master to agree on the speed of light – both of them thinking that the others clocks are slow but their rulers are short. But, when Ludvig looks down the well at his brother measuring the speed of light, he sees Albert's clocks running slow but his ruler the correct length. He has to conclude therefore that, down there, light genuinely does travel more slowly than it does for him. In short, we can say that, from the point of view of an observer at the top of the well, the speed of light at the bottom of a potential well of depth $\Delta\varphi$ is

$$c\sqrt{1 \ - \ 2\Delta\phi/c^2}$$

It will not surprise you to learn that if the potential well is deep enough, the speed of light as viewed from someone a great distance away becomes zero and light will never 'escape' from such a well. (Do not forget, however, that from the point of view of someone down inside the well, light travels at its normal speed.)

Nor will it surprise you to learn that this is the real reason why light can never escape from a Black Hole. From the point of view of a distant outside observer, at the surface of a Black Hole time stops and light never moves!

The Bending of Light (continued)

We are now in a position to better understand why gravity bends light.

When Betty described the way light bends in an accelerating lift, she explained that it only appears to bend because, during the time it takes for the light to cross the lift, the lift has accelerated and moved an extra distance $\frac{1}{2}at^2$ (see page 103). This is true, but it doesn't really explain why *gravity* bends light. The Principle of Equivalence says that it must – but this isn't really an explanation either.

The answer lies in the fact that light travels more slowly at the bottom of a well than at the top. Let Betty explain.

'I have have just realised something important.'

'Oh. What's that?' said Arthur

'I have just realised why gravity bends light.'

'That's amazing. Go on – show me.'

You know it is often said that sound travels well over water and that you can sometimes hear two people having a conversation on the other side of a lake as clearly as if they were next to you.'

'Yes. I know that.'

'Well, the reason is that the water immediately over the lake is cooler than the water higher up and that the speed of sound is less in cool air because the air molecules are not moving so quickly.'

'That sounds as if it should take longer for the sound to cross the water than before and the conversation should be less audible, not more.'

'No that's not what happens. The important thing is that the wavefronts closer to the water get dragged behind, rather like a line of soldiers on parade where one end of the row is hampered by some boggy ground. Here, I will show you.'

Whereupon Betty sketched out a drawing a bit like this:

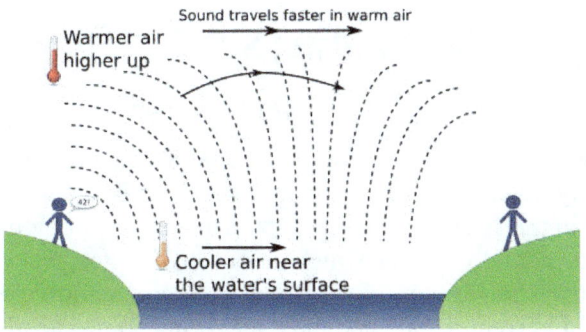

Sound travels faster in warm air

Warmer air higher up

Cooler air near the water's surface

Source: inspiringscience.net

'See how the waves curl over the water, bending as they go, focussing themselves on the opposite shore.' explained Betty.

'I see that. And I suppose it is the same with light in a gravitational field. The light at the bottom travels more slowly than the light at the top so it bends over, just as if it was pulled down by gravity.'

'Exactly so.'

'Does it check out? I mean, can you show that the amount of bending in a 1G gravitational field is exactly the same as the bending you get in a lift accelerating at 1G?'

'I think I would bet my life on it.' said Betty.

Betty is safe. It does indeed check out. Once we take into account the effects of Time Dilation, Maxwell can agree with Newton. The latter can maintain that it is just the effect of the force of gravity on his stream of corpuscles; the former can claim that the bending is due to the retardation of his electromagnetic wavefronts. In truth, light is just doing what it has to do given the constraints imposed upon it by the rules of Relativity.

'But there is still something I don't understand.' said Arthur.

'What's that?'

'That factor of 2 you mentioned when we were talking about the Schwartzchild Radius of a black hole' (page 107)

'Yes, we haven't got to the bottom of that yet, have we?'

The Rotating Space Station

The Ehrenfest Paradox

The paradoxes we have discussed so far are, these days, beyond controversy. Even the Twins Paradox, which caused a good deal of argument in the 1960's, is now accepted as fact – though there are still books in print and websites which do not give the correct explanation of the paradox. The paradox of the Rotating Space Station, however, is still a subject of current debate and may yet spring some surprises on us.

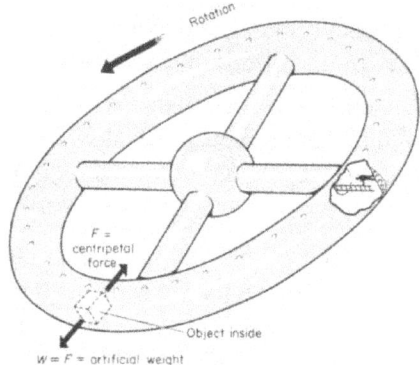

Rotating Space Station
Source:https://astarmathsandphysics.com/

A popular idea in science fiction is the concept of a space station in the form of a rotating wheel. Objects and people in the rim of the wheel will experience a centripetal acceleration towards the centre and will effectively imagine that they are living in a world with artificial gravity. If the rim is travelling at a speed v and the radius of the wheel is R then the acceleration is given by the formula v^2/R For a station of radius 90 m, the speed necessary to provide 1G gravity would be 30 ms^{-1}and it would rotate once every 19 seconds.

The paradox is this. If Albert was to take a metre ruler aboard the station and carefully measure the radius, he would find that it is exactly 90 m. There is no reason to expect the ruler to be contracted because it is moving at right angles to its length. In any case, if it was contracted, he would still find that 90 rulers would fit along the radial corridor because the space station would also be contracted. Now it would appear that the same argument applies to a measurement of the circumference. If the ruler undergoes length contraction, so does the space station and Albert will still always measure the circumference to be $2\pi \times 90 = 565.5$ m long.

But what about his brother Ludvig, hovering over the docking port, watching the station rotating below him? Suppose that there is some stationary scaffolding, perhaps left over from the construction of the station, which exactly frames the now rotating station. Ludvig can check the dimensions of the radius and circumference directly. There are only two possibilities: either the station still fits the scaffolding exactly (i.e. it has the same dimensions that it had before it was set into rotation) or its dimensions have changed. When this idea was was first formulated by Paul Ehrenfest in 1909 Sir Arthur Eddington thought that, since the moving rim would have to be length contracted, the radius would contract as well as the circumference. In effect he was putting forward the idea that gravity – or, in this case, artificial gravity – would cause a length contraction as well as a time dilation effect. There is no evidence that this is the case ,so we will assume that, to Ludvig, the space station retains the same dimensions all the time.

But Ludvig now has to face the fact that the circumference is moving at 30 ms^{-1} and any rulers and objects, including the fabric of the space station itself ought to be length contracted by a factor k. If he and Albert were to compare metre rulers as they passed each other, then. like the station master and the guard, they would each conclude that the other's rulers were length contracted.

Ludvig now watches Albert lay out a series of metre rulers round the circumference of the station. Albert can see no reason to suppose that he will need any more than the expected 565.5 m; Ludvig predicts that he will need a bit more[34]. Who is right? Is there a contraction or isn't there?

And what about Albert's view? Surely, he can argue that it is not the station which is rotating, it is the scaffolding; in which case it is the latter which should be contracted, not the former.

This issue raises deep philosophical questions about the nature of linear and rotational motion and their relation with Space and Time.

For Newton, Space and Time were absolute. He was perfectly

---

34 Only 5.6 thousandths of a millionth of a millimetre actually!

comfortable with the idea that the Sun was the centre of the universe and was absolutely stationary and that the Earth rotated round the Sun once a year, not relative to anything else but just through stationary space. But Galileo had shown that the laws which governed the behaviour of falling bodies was independent of motion when he pointed out that a cannon ball, dropped from the mast of a moving ship, would still fall at the base of the mast. At the very least this suggested that motion was a relative rather than an absolute concept. So Newton looked for evidence that Space and Time were absolute – and he thought he had found it in the behaviour of rotating bodies.

In the Scholium on Space and Time in the Introduction to Book I of the Principia he says:

> *We have some arguments to guide us, partly from the apparent motions, which are the differences of the true motions; partly from the forces, which are the causes and effects of the true motions. For instance, if two globes kept at a given distance one from the other, by means of a cord that connects them, were revolved about their common centre of gravity; we might, from the tension of the cord, discover the endeavour of the globes to recede from the axis of their motion. ... And thus we might find both the quantity and the determination of this circular motion, even in an immense vacuum, where there was nothing external or sensible with which the globes could be compared.*

There are two things to say about this argument. The first is that it is an argument for the existence of absolute *rotation*, not absolute *motion*. In fact we now know that there is no way absolute linear motion can be detected; indeed, this principle is at the heart of Special Relativity and is the reason why experiments such as the Michelson-Morley experiment singularly failed to detect the motion of the Earth through space.

As an argument for the existence of absolute rotation it carries more weight. Essentially it is saying that if the occupants of the space station experience artificial gravity, then they can infer that the space station is rotating without reference to anything outside the station.

This argument was challenged by Enrst Mach who pointed out that the argument doesn't work in a completely empty universe. Let me explain. Suppose we have finished constructing the space station out of all the material in the universe so that our starting point is an empty universe with a non-rotating space station with no artificial gravity. How are we going to start the space station rotating? Because of the conservation of angular momentum, the only way is to get something else to rotate in the opposite direction. The usual technique it to attach rockets to the rim and fire them up. Now, rockets work by ejecting mass at high speed out of the exhaust so at the end of the day, when the space station is rotating up to speed, we have a space station rotating one way – and a whole load of exhaust gases rotating in the opposite direction. In other words, the universe is no longer empty and the rotation of the station can be measured relative to the cloud of exhaust gas.

Notwithstanding Mach's philosophical objection, it is generally assumed these days that Newton was right and that in practice we can determine the rotation of an isolated object with reference to the distant galaxies. (It turns out we can even talk about absolute motion through the cosmos too as we shall see in a later chapter.)

Returning now to the question of Albert's measurement of the circumference, there are still some authors who think that Albert will only need the expected 565.5 m. They usually explain the apparent lack of any length contraction effect by appealing to the idea that Albert and Ludvig cannot compare rulers as they pass by in exactly the same way that the guard and the station master do, because Albert is not in an inertial frame and cannot construct a system of synchronised clocks around the circumference of the space station in any consistent way. It is therefore incorrect to apply a simple length contraction effect to the rulers on the circumference.

I do not see the force of this objection at all. The existence or otherwise of the length contraction effect does not depend on the radius of the space station and, by making the radius as large as you please, you can make Albert's frame of reference as nearly inertial as you please too. It is my belief, therefore, that when Albert comes to lay out his rulers, he will find that he is (a few billionths of a millimetre!) short.

This is bound to cause him some surprise. He has measured the radius of the station and the circumference – but found that the ratio is slightly *more* than 2π. Nevertheless, I believe that he must accept this as fact. In a rotating frame of reference the circumference of a circle is greater than 2π times its radius. In short, in a rotating frame of reference, *space is not Euclidean.*

Although we are not generally aware of the the fact, we humans on planet Earth already live in a non-Euclidean world. If you draw a circle round London on a globe passing through, say, New York and then measure the length of its circumference you will find that is is always *less* than 2π times the radius of the circle (measured along the surface of the Earth, of course).

If we lived on the surface of a Pringle (a crisp shaped like a saddle) you would find, as Albert does, that the length of the circumference of a circle is *greater* than 2π times the radius.

This was the view adopted by Einstein himself who wrote in 1919

> *One must take into account that a rigid circular disk at rest would have to snap when set into rotation, because of the shortening of the tangential fibres and the non-shortening of the radial ones. Similarly, a rigid disk in rotation (made by casting) would have to shatter as a result of the inverse changes in length if one attempts to bring it to the state of rest. If you take these facts fully into consideration, your paradox disappears'.*

When Einstein realised this, he also realised that his Principle of Equivalence meant that any gravitational field would alter the geometry of space round it too. Albert lives in a world in which his artificial gravity increases with distance from the centre of rotation and is directed radially *outwards*. He has also discovered that the distance round the circumference of a circle is *more* than 2π times the radius of the circle . On the other hand, the gravitational field round a massive object decreases according to an inverse square law and is directed radially *inwards*. The implication of this is that the distance round the circumference of a circle round a massive object ought to be *less* than 2π times the radius of the circle and that the discrepancy increases as the radius gets smaller.[35] It is essentially this fact which causes the orbit of Mercury to behave in a way which is different from the predictions of Newtonian gravity.

So what exactly does happen when the space station is spun up to speed? As far as Ludvig is concerned, the radius and the length of the actual circumference of the station remain unaltered but the girders which make up the circumference are length contracted. These girders are therefore placed under stress and will stretch to accommodate the extra length required[36]. This stress is totally independent of (and very much smaller than) the stress which any rotating object with mass will normally experience. You would get the same effect if you were to make a flat model of the space station out of rubber and then press it onto a Pringle.

Ludvig will also notice that a clock at the hub of the space station keeps the same time as his own clock but the clocks on the rim will suffer from time dilation and will run slow.

Albert, however, will have a different explanation for the same effects. He will regard himself as stationary but he will experience a regime of artificial gravity which increases with distance from the central hub of the station. Since he has to climb up towards the hub against this gravity, clocks at the rim will run more slowly than the clock at the centre. His measurement of the length of the radius and the circumference will reveal to him that the space he inhabits is not Euclidean but he will interpret this as being due to the presence of the gravitational field. Either way, two measurable and physical phenomena will occur. a) stresses will exist in the structure of the station and b) Albert will age more slowly than Ludvig.

35 It is not generally possible to *define* let alone *measure* the radius of a circular ring around a massive object because of the distortion of space that it produces. It is, however, possible in principle to measure the difference in circumference of two rings round a such an object and the difference will be found to be less than 2π times the separation of the rings.
36 In all probability the stress in the rim will compress the radial components and the whole space station will actually shrink very slightly as well.

When Betty was explaining the reason why clocks at the back of an accelerating space ship run more slowly than clocks at the front she was unable to derive a totally accurate expression (page 110) because the acceleration necessarily causes the *speed* to change and if the ship is moving at a relativistic speed, you cannot simply say that the speed increases by a (the acceleration) every second. In the case of the rotating space station, however, the acceleration only causes a change in *direction*, not speed. This makes it possible to derive an accurate expression for gravitational time dilation.

Since the rim of the station is rotating at a speed ωR where ω is the angular velocity of the station, a clock on the rim must be running slower than a clock at the centre by a factor of

$$\frac{1}{\sqrt{1 - \omega^2 R^2 / c^2}}$$

Now, in Albert's rotating frame, the artificial gravity is directed outwards and is proportional to the distance r from the centre. In fact $g = \omega^2 r$ where ω is the angular velocity of the station.

The difference in gravitational potential between the centre and the rim is therefore

$$\Delta\phi = \int_0^R \omega^2 r \, dr = \tfrac{1}{2}\, \omega^2 R^2$$

with the potential greater at the centre than at the edge.

Eliminating $\omega^2 R^2$ from these two equations we get the correct equation for the gravitational time dilation factor between two points whose gravitational potential differs by $\Delta\varphi$:

$$\frac{1}{\sqrt{1 - 2\Delta\phi / c^2}}$$

Inside a Black Hole

'What would it actually be like to fall into a Black Hole?,' asked Arthur one day.

'That greatly depends on the mass of the hole.' answered Betty. 'If the Earth was compressed in to the size of a marble, it would become a black hole, but you wouldn't necessarily fall into it. You could orbit around it in your space ship just like orbiting the Earth in the space shuttle.'

'Cool. But what would you *see*?'

'Not a lot. Could you see a black marble at a distance of 4000 miles?'

'No. But this is a *Black Hole* we are talking about. Wouldn't it have any visible effects at all?'

'Yes it would – if you looked at it through a telescope you would see that the stars behind it were sort of ' repelled' from it due to the bending of starlight as it passed close by. You might even see a complete circle of light round it known as an Einstein Ring if there was a bright object exactly behind it like this:'

J095629.77+510006.6

Source: Hubble Space Telescope

'That's cool,' said Arthur, " but what would the black hole itself look like?'

'Suppose you fired a cannister out of the back of the space ship at exactly the right speed,' suggested Betty. 'The cannister would now be left behind effectively stationary and you could observe it through a telescope as it fell towards the black hole. Obviously it would fall faster and faster until it was swallowed up by it – but this is not what you would actually *see*. Suppose also that the cannister emitted a flash of light every second: due to gravitational time dilation, as it approached what is called the 'event horizon'[37] of the black hole, the light would get redder and redder and the intervals between the flashes longer and longer – so long, in fact that you would never actually see it enter the hole, it would just fade from sight. In fact, if you could somehow switch off the red shift, what you would actually see on the 'surface' of the black hole would be the remains of everything that had ever fallen into it!'

'You mean like some sort of cosmic rubbish tip?'

'Exactly so. Every cannister, every shopping trolley, every asteroid everything that ever fell into it sort of plastered onto its surface!'

'That's amazing! But I thought black holes swallowed up everything in the vicinity.'

'Yes they do. Your cannister does fall into the black hole – it is only from your position outside the hole that you can still see it falling. You could say that, from your point of view, the inside of the black hole doesn't exist. It really is a hole in your universe and from your point of view things never actually leave your universe, they just fade away as they approach the event horizon. At the event horizon time stops and the distortion of space becomes infinite'

'Is that what they mean by a 'spacetime singularity'? A place where times stops and space becomes infinite?' asked Arthur.

'Actually that is a very good question and even Einstein was confused by it for several decades. No, the event horizon is not where the singularity is. The singularity is at the centre of the black hole. There the laws of physics completely break down and pretty well everything becomes infinite.'

'But it sounds to me as if things get pretty extreme at the event horizon.' said Arthur..

---

37 The event horizon is the surface of a sphere round the black hole of radius equal to the Schwartzchild radius R_{sch}. See page 107. More precisely, since the idea of a radius is problematical owing to the distortion of space near the black hole, it is better described as the surface of a sphere which has an area of $4\pi R_{sch}^2$.

'No. That's only how it looks to someone outside. If you were inside the cannister you wouldn't notice the event horizon at all. Everything would appear perfectly normal – well, at least, almost normal.'

'What do you mean?' asked Arthur.

'Well, since the cannister is in free fall, the person inside wouldn't feel any gravity. Like an astronaut in orbit, he would be weightless. But, as he got closer to the black hole he would start to feel what are called 'tidal forces' on him. Those parts of his body which are closer to the hole would be attracted more strongly than those parts of his body that are further away. In fact he would be first pulled into a vertical position and then stretched like a martyr on the rack.'

'How horrible! How close could he get to the hole without being torn apart?'

'That depends on the mass of the black hole. In the case of a black hole with a mass equal to that of the Earth things would start to get pretty uncomfortable at a distance of about 60 km – but then he wouldn't have long to live at that point!'

'So you could never get to the event horizon because you would be dead before you got there!'

'Not necessarily. If the black hole was really big, the tidal forces are relatively small and it would be perfectly possible to take a space ship right down to and inside the event horizon. In fact, if the black hole was solid and sufficiently massive you could land your space ship and walk around on its surface.'[38]

'Really?'

'Yes. You remember I said that the Schwarzchild radius was $R_{sch} = \dfrac{2GM}{c^2}$ ' (see page 107).

'Yes – and I queried that factor of 2 as well, didn't I?'

'Yes, you did. But leaving that aside for the moment, the gravitational field strength at the surface of a planet or star is $g_s = \dfrac{GM}{R_s^2}$. If we put

---

38 Betty is stretching a point here. Since it would take an infinite amount of energy to take off from the 'surface' of a black hole, it would take an infinite amount of rocket fuel to decelerate a rocket and land it.

$R_s = R_{sch}$ and eliminate GM we find that $R_{sch} = \dfrac{c^2}{2g}$. Now putting $g = 10$ ms^{-2} and $c = 3 \times 10^8$ ms^{-1} we deduce that the radius of the black hole which has a surface gravity of 1G will be about 4.5×10^{12} km or a little bit less than half a light year.'

'That sounds pretty big!'

'Yes, it is. And it turns out to have a mass of about twice the mass of the whole Milky Way too – but who is to say that such an object does not exist somewhere in the universe?'

'That's pretty cool! So what would it be like to walk on the surface of this black hole. You said that gravity at the surface would be 1G so it would be just like walking round on Earth. Right?'

'Correct. But it might not be a pleasant place to be all the same.'

'Why not? I suppose all sorts of weird things happen in a place where time has stopped and where the distortion of space is infinite.'

'No. You still haven't understood the nature of the event horizon.' said Betty. 'Time stops and space is infinitely distorted only from the point of view of someone *outside* the system. As far as you are concerned, time flows at its normal rate and the space around you looks perfectly normal. The event horizon is what is known as a 'coordinate singularity' because it only looks like a singularity from certain points of view.'

'You mentioned the singularity at the centre of the black hole. Is that a coordinate singularity too?' asked Arthur.

'Definitely not. That is a real physical singularity and exists whichever way you look at it.'

'OK, ' said Arthur. 'So why would walking around on the surface of this black hole be dangerous?'

'Well all the light from the other galaxies in the universe would be bent so that it fell vertically down; it would also be massively blue-shifted so it would be like standing under a gamma ray spotlight.'

'Also,' she continued, 'the black hole would probably be sucking in gaseous material from the vicinity, all of which would be moving at near the speed of light carrying massive amounts of energy. No. I don't think you could survive. In any case, I don't think there would be an actual surface to walk on.'

'Why not?'

'Because, in spite of the huge mass of this object, its density would be quite low – about 5 milligrams per cubic metre.'

'So I would just go on falling, would I?'

'Yes – until the tidal forces became too great and you were stretched out like a piece of spaghetti.'

'So what would it look like – actually being *inside* a black hole?' Arthur wondered.

'Frankly nobody knows. And if you went there, you would never be able to come back and tell us all about it.'

'Pity. Could it be an entry point into another universe?'

'Well a number of highly respected scientists as well as science fiction novelists have explored that idea but, to be honest, your guess is a good as theirs. Still, it makes a good story.'

The Bending of Starlight

'I think I have got a bit further with the bending of starlight problem,' said Betty a few days later. 'I asked a question in a physics forum and got a lot of responses from people who obviously know a lot about General Relativity.'

'That's great,' said Arthur. 'What did they say?'

'Basically they said that if I wanted to understand it, I would have to take a post-graduate degree course in General Relativity.'

'That's not very helpful, is it?'

'No. But one of the contributors was a bit more helpful. He (or she) said that the bending of light in a uniform field was a local effect and was correctly explained by the Equivalence Principle, but the other half of the bending was a global effect due to the fact that, as he put it 'A line of little labs[39] along the light path won't quite fit together the way Euclid says they would''

'What does that mean?'

'I take it to mean that the presence of a massive object distorts space in such a way that, although locally the bending in each little 'lab' was due solely to the Equivalence Principle, when you put lots of little 'labs' side by side, they don't align properly and you end up with more bending than you expect.'

'Yes, I think I see what he might mean. After all, suppose you made a complete set of one inch maps of the whole Earth. Each individual map would look perfectly square and reasonable – but if you tried to stitch them together, you would soon find that the edges wouldn't line up properly because the Earth surface is not flat,' said Arthur.

'I agree. And they weren't very happy about my suggestion that you could explain the ordinary 'Newtonian' bending by using the idea that light travels more slowly at the bottom of a well than at the top.'

'Why not?'

'Well, there was a lot of mention of coordinate systems and the fact that it is much more difficult to define a coordinate system in the

---

39 By 'labs' the author of the quote is referring to small local laboratories in which experiments on the bending of light etc. can be carried out.

presence of a gravitating mass in which phrases like 'the speed of light at the bottom of the well' have any physical meaning.'

'But this is splitting hairs, isn't it?' Arthur objected, 'Surely if the idea that the speed of light varies down the well gives exactly the right result, there must be some truth in it, mustn't there?'

'It wouldn't be the first incorrect theory to give the right result. Back in 1783 a clergyman called John Michell worked out the Schwartzchild radius of a black hole. It was well known that if you gave a projectile enough kinetic energy it could completely escape from the surface of a planet or star. In fact the 'escape velocity' of a star or planet could be calculated from the expression

$$\frac{1}{2}mv^2 = \frac{GMm}{R}$$

By putting $v = c$ and rearranging we get the correct expression for the Schwartzchild radius of a black hole:

$$R_{sch} = \frac{2\,GM}{c^2}$$

which even includes the factor of 2 which we are so bothered about!'[40]

'I see what you mean. But surely we can say that the bending of light round a star is *analogous* to the bending of sound waves over a lake, can't we?'

'Personally, I don't see the harm in that, provided we bear in mind that at the end of the day, it is only an analogy and that its scope is strictly limited.'

'Is there any way we can extend the analogy to cover the spatial distortion as well?'

'I have been thinking about that and I think there is' said Betty. 'You know you said that the one inch maps would not fit together because the surface of the Earth was flat. Well, if you went ahead and stitched them together anyway, you would, of course, end up with a pretty good approximation to a globe. It turns out that if you stitch together all the 'little labs' around a gravitating mass you end up with a surface that looks a bit like a volcano – like this':[41]

---

40 For a comment about the derivation of this formula see the appendix.
41 This surface is known as Flamm's Paraboloid and it is the surface of revolution of the curve $z = -\sqrt{R - R_{sch}}$ where R_{sch} is the Schwartzchild

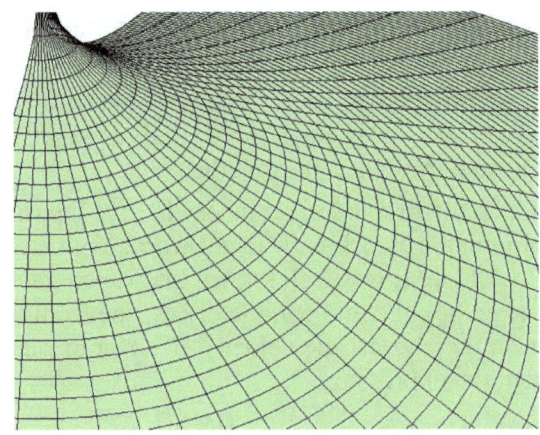

'If you look at the concentric circles, the circumference of each circle differs from the next by the same amount, but as you get closer to the origin, the *radial* separation along the surface gets greater and greater. Effectively what this means is that the closer you get to a massive object, the further you have to travel to make progress.'

'Is that why to me outside a black hole, it would seem to take for ever for my cannister to fall into the hole?' asked Arthur.

'I am sure my physics mentor would run a mile if he heard you say that, but it seems to make sense to me. At least, I think it would be possible to choose a coordinate system in which it makes sense.'

'I think I am beginning to see what you are getting at. Do you mean that light which travels close to the massive object has somehow got further to travel?'

'Yes – that is exactly what I mean. You know that old analogy with soldiers on parade which explains why light bends at an oblique surface. I think we can usefully use the same trick here.

'Suppose a column of soldiers is marching towards us across the shoulder of this volcano-like surface. The soldiers all march at exactly the same speed, but the soldiers who have to march closer to the origin have to climb over a hill and therefore have to march further – like this':

---

radius.

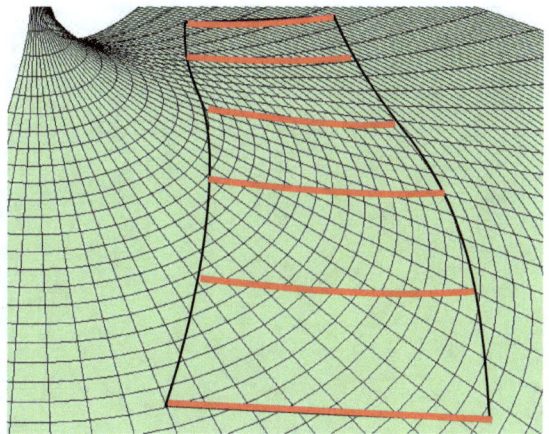

'Naturally these soldiers are somewhat held back and the line wheels round.'

'That's a great picture!' exclaimed Arthur, 'I don't care if it isn't strictly accurate, it really does enable you to visualize how the spatial distortions of the 'little labs' which your mentor spoke about fit together in a non-Euclidean way to cause the light to bend.'

'Yes, I agree. We can also include the ordinary 'Newtonian' effects as well by supposing the the ground gets more and more boggy, the higher up the volcano you go. This will cause the soldiers to slow down and cause the line to bend even more.'

'Exactly twice as much I suppose,' said Arthur.

'Well I wouldn't go so far as to say that the idea can be made quantitative but I do think the analogy is a pretty good one.

'Here is a picture showing both the spatial ('Einsteinian') and temporal ('Newtonian') effects at the same time,' said Betty:

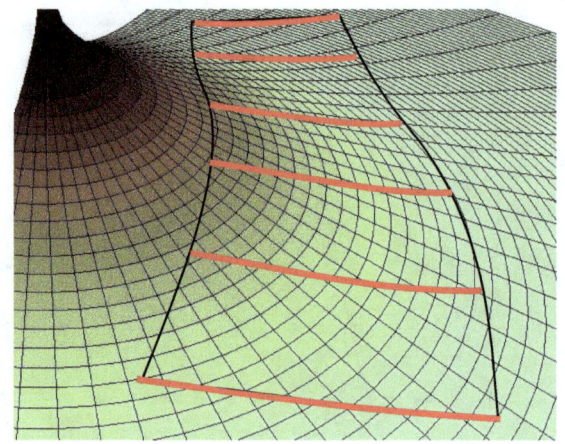

'One interesting fact about this picture is that, when the line of soldiers points directly up and down the slope (i.e. radially), all the soldiers are marching horizontally. None of them are climbing up or down. The only thing, therefore, which causes them to bend is the boggy ground effect. In the context of starlight passing a star, when the light is travelling at right angles to the gravitational field (i.e. when it is at its closest to the star), only the 'Newtonian' or Equivalence Principle effect occurs. The bending due to spatial distortion only occurs when there is a component of the gravitational field in the direction of travel of the light. This explains why you only need to consider the Equivalence Principle when considering the bending of light in a lift, or indeed, when standing on the ground.'

'I am pretty impressed with your idea,' said Arthur,' but I am a bit confused by one thing.''What's that?' asked Betty.

'Well, you often see the bending of starlight explained by using the analogy of a rubber sheet weighted down by a heavy weight. This is supposed to illustrate how the presence of a massive object distorts space around it. Marbles projected towards the weight appear to bend just like light and can even be made to orbit the weight just like a satellite. Here is a typical picture:'

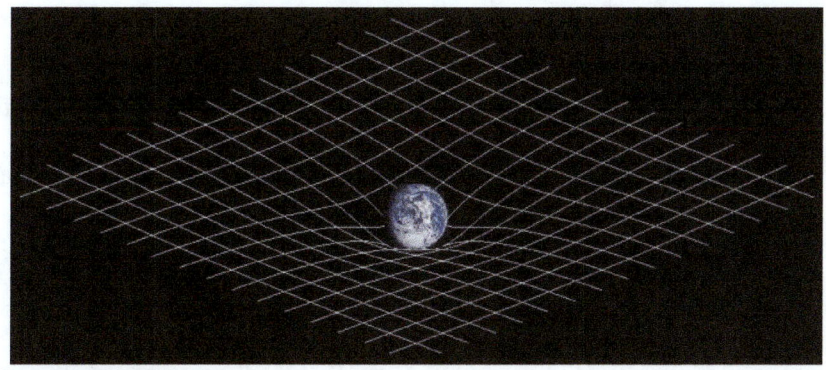

Source: Wikipedia

'What I want to know is: is this the same analogy or is it something different?'

'My view,' said Betty, ' is that it is a terrible analogy and has nothing to do with the distortion of spacetime at all. It does, however, model the classical situation reasonably well. The distortion of the weight generates an approximation to what is known as a gravitational potential well and when a marble rolls past the weight it is acted upon by a sideways force which depends on the slope of the well.'

'So if it doesn't model the distortion of space, why do people go on using it when trying to explain the bending of starlight?'

'Habit, I suppose. My volcano analogy has nothing to do with potential wells. For a start, the shape of the volcano is a paraboloid[42] whereas the shape of a potential well is a hyperboloid[43]. More importantly, it genuinely does represent the way in which space is distorted round a massive object.'

'Well, I certainly don't think I am going to get a better understanding of why Einstein's prediction of the bending of starlight is twice what you would expect. Not without doing a graduate course in General Relativity, at any rate!' said Arthur, gratefully.

---

42 It is the surface of revolution formed by rotating the curve $y = -\sqrt{x-1}$ about the y axis and is known as Flamm's paraboloid.

43 The surface of revolution formed by rotating the curve $y = -1/x$ about the y axis.

Geodesics in spacetime

Everybody knows that 'light travels in straight lines' – but in fact this statement is both tautological (i.e. true by definition) and manifestly false. Of course, it all depends on what you mean by a 'straight line' and, as Betty's mentor was at pains to point out, that all depends on what coordinate system you adopt. In General Relativity, choosing the correct coordinate system is very difficult, so we can only be sure of getting the right answers if we choose to employ only those quantities which are independent of the coordinate system. The most important of these is the quantity we met earlier (see page 57) called the 'interval' between two events which was defined as:

$$I = \sqrt{\Delta t^2 - (\Delta x^2 + \Delta y^2 + \Delta z^2)/c^2}$$

and which is a measure of the 'distance' between and event (t, x, y, z) and the event ($t + \Delta t$, $x + \Delta x$, $y + \Delta y$, $z + \Delta z$) . When an object such as a light beam or a satellite makes its way through spacetime, if you add up all the little intervals along the way, you end up with what is called the *'proper time'* for the journey.

For example: you will recall that Albert reached Alpha Centauri in only 3 years (because the distance to the start was shrunk by length contraction) (see page 23). It also took him 3 years to get back. The total time for the complete journey was 6 years. This is the *proper time* for the journey.

When he got back he found that his stay-at-home brother had aged not 6 but 10 years. This is perfectly consistent because, in Albert's frame of reference, his arrival at Alpha Centauri had coordinates (3, 0) while in Ludvig's frame, the same event had coordinates (5, 4). the *interval* between these two events (Albert's departure and arrival) is calculated by Albert to be $\sqrt{(9 - 0)} = 3$ years while Ludvig calculates it to be $\sqrt{(25 - 16)}$ which is 3 years also.

Actually, Albert didn't really need to go to Alpha Centauri at all. Just flying at 60% of the speed of light would have been sufficient to make his clocks run slow. By the same token, he could – in principle – have climbed down a deep well for a while to get the same effect!

The point of all this is to demonstrate the counter-intuitive fact that objects in free fall (like Ludvig) undergo a *larger* proper time interval than objects which take a more roundabout route through spacetime. In fact this is a fundamental principle in General Relativity – any object which is in 'free fall' (i.e. allowed to move entirely according to the local shape of spacetime) follows what is called a *geodesic* which *maximises* the proper time along the line.

Now, we are reasonably familiar with the idea of geodesics on Earth as being the shortest route which an aircraft might take between two cities and the line which a string will adopt when stretched across a globe. But the idea that a geodesic in spacetime is a line which *maximises* the proper time between two events in spacetime is, to say the least, a little strange. It is all to do with that minus sign in the expression for the interval – and the vital fact that you cannot exceed the speed of light[44]. The condition implies that *at every stage on the journey* the Δt^2 term is always larger than the ($\Delta x^2 + \Delta y^2 + \Delta z^2$)/$c^2$ term. Taking extra excursions into the x, y and z directions along the way is only going to *reduce* the overall proper time.

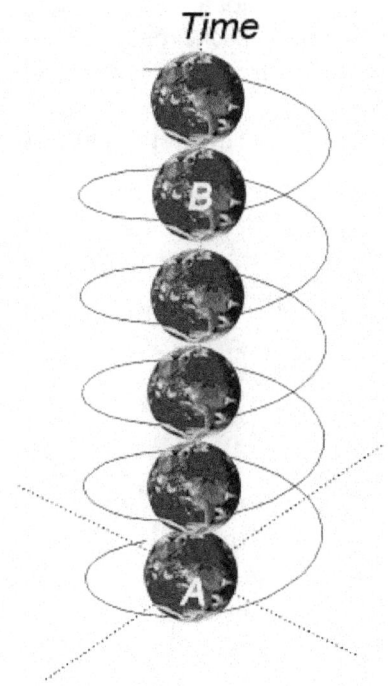

Time

Consider the case of a satellite orbiting Earth. Since it orbits in a plane we can ignore the z coordinate and just plot its x and y position horizontally and time vertically as in the accompanying diagram. The satellite is following a geodesic route from A to B.

But why does it choose to circle the planet four times? Why doesn't it just go straight from A to B? The answer is that, if it did, it would stay in the same place all the time

---

44 Whether a geodesic is a maximum or a minimum path is not really relevant. What is important is that small *deviations* from the path make very little difference to the distance or proper time. This is true at maximum, minimum and stationary points.

and the proper time interval between the two events would simply be T (the time between the two events). But, by taking a more roundabout route, the satellite can change the proper time interval between A and B. If the satellite takes an excursion to a Mars and back, then, like Albert, its proper time will *decrease*. But it is not necessarily the case that the proper time will always decrease. If the satellite were to climb up a bit out of the planet's gravity, its clocks would speed up; then, after spending a little while there, it could come down again to point B with its clocks reading *more* than T. But the satellite cannot go too far or too fast as this would cause its clocks to slow down again. There will, in fact, be an optimum trajectory which takes the satellite out and back such that the proper time, as recorded by its on board clocks, is the *maximum possible*. The route described here is not an orbital route, it is what might be described as a 'ballistic' (out and back) route. It turns out, however, that there are many ways of getting from A to B along a trajectory whose proper time is a maximum, just as there is often more than one way of traversing a mountainous landscape in the shortest possible distance.

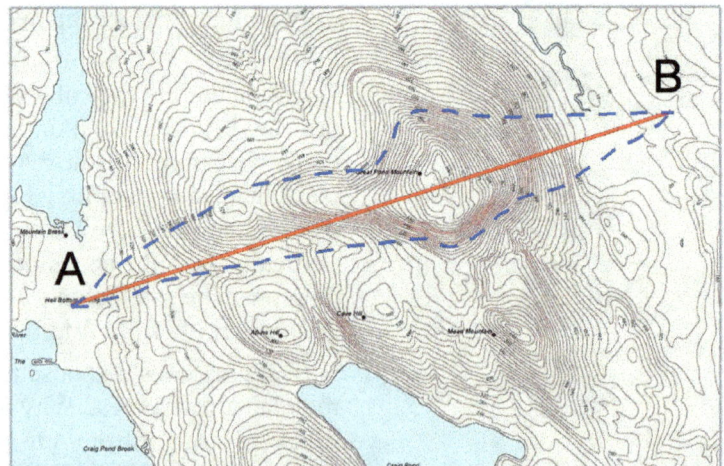

Suppose, for example, you wish to walk from A to B across the map shown above. The direct route would take you right over the top of the mountain and, quite apart from the extra effort involved in climbing it, the distance walked would be greater than the straight line distance between A and B. The two blue routes are shorter because although they wiggle sideways, they do not go up and down so much.

Each blue route is the *shortest possible route on that side of the mountain*. The vital point here is that small deviations from the route will always result in a longer journey. Each route is a *local minimum*. True, one route is probably longer than the other – but both routes are in a sense the best possible way of getting from A to B.

Similarly, the satellite has many ways for getting from A to B and by choosing a route which goes *round the back of the Earth* one or more times, the satellite can maximise the proper time. Some of these orbits will be elliptical; the one illustrated with four circuits of the Earth is circular. Locally each route is a maximum and therefore each is a perfectly valid geodesic between A and B. (Which one the satellite actually takes depends on the speed and direction with which the satellite is launched at A.)

The geodesic which is followed by light is special because the proper time interval between any two events along the path of a ray of light is precisely zero. If a light ray starts off at (0, 0) it will reach a point at a distance x in a time $t = x/c$. The interval between (0, 0) and (x, x/c) is $\sqrt{x^2/c^2 - (x^2 + 0 + 0)/c^2} = 0$. We cannot therefore use the 'maximise' rule to determine the path taken by a ray of light. But there again, we do not have to. One of the fundamental principles of Relativity is that the speed of light is constant. All light-like geodesics, therefore, point at 45° to the temporal axis by definition.

Part 4: Cosmology

The Expanding Universe

We have known for a long time that the universe is expanding. Virtually all the galaxies which we can see beyond our own local group (which includes the Andromeda galaxy) show a red shift implying that they are receding from us and initial measurements suggested a simple proportional relationship between recessional velocity and distance.

In 2012 NASA released an image – the Hubble Extreme Deep Field – of an apparently completely empty area of the sky which was the result of many hundreds of 20 minute exposures taken over a period of 10 years.

Hubble Extreme Deep Field - 2012

The image is about 2 arc minutes across – about the size of a large crater on the Moon. It shows over 5000 galaxies but how far away are they and how fast are they moving?

The first problem is that some of the small faint ones may, in fact, be relatively close. It is almost impossible to measure the red shift of these galaxies directly (a process which requires identifying specific spectral lines in the spectrum of the light) as they are incredibly faint. Fortunately, the image above is a composite of several images taken using different filters, one of which was an ultra violet filter. Any reasonably close galaxy will show some emission in the UV part of the spectrum but a very distant, highly red shifted galaxy will not show any UV. Once you have weeded out the nearer ones, a simple measurement of the size and brightness of the remaining ones will give an estimate of their distance. Some of them, it is believed, are seen as they were less than a billion years after the big bang.

The standard method of measuring the distance to a distant galaxy is to look for a supernova in it and measure its brightness. It is clearly impossible to do this with any of the galaxies in the above image but the distances to hundreds of galaxies with red shifts[45] of up to 2.5 have been measured by this method.

Another method of measuring the distance to a distant galaxy is to monitor the incredibly violent bursts of gamma rays which are emitted when a star suddenly collapses into a neutron star or a black hole. The galaxy from which the burst came from can then be studied in detail and some of them have been found to have red shifts of up to 8 or even more.

The question is – does the simple proportional relationship between recessional velocity and distance extend out this far?

The first problem which has to be addressed is the simple fact that we do not see distant galaxies as they are *now* but as they were when they emitted the light which we can see. This makes things very confusing. Fortunately, however, there are a number of things which we do *not* have to worry about. Firstly, we do not have to worry about space-time curvature. As far as we can tell, our universe is spatially *flat* [46](hooray!). This means that two light beams which set out parallel (or

---

45 Throughout this book I shall use the physical definition of red shift $Z = \lambda/\lambda_0$. Astronomers generally use the definition $z = \Delta\lambda/\lambda_0$. To convert one to the other use $Z = z + 1$.

46 There has been much popular speculation about spatially 'closed' universes which are finite like the surface of a sphere or temporally 'closed' universes

indeed at any angle) will remain parallel (or at the same angle) for ever.

Secondly we do not have to worry about length contraction or time dilation (at least, not much). This is because there is an important sense in which, on a cosmological scale, the galaxies can all be regarded as being stationary.

Thirdly, we can, without assuming anything specific, place ourselves at the origin because, if the universe is the same everywhere, any place can be regarded as an origin.

So let us start our exploration of the possible expanding universes with the classic picture of a possible universe – one which is infinite in extent and absolutely the same everywhere. Its geometry is strictly Euclidean and we can imagine it to be populated with an infinite number of identical galaxies all exactly the same distance apart. Something like this:

If this universe is static a serious problem arises. In whatever direction you look you ought to see a galaxy and the night sky should be as bright as the sun. But, if the universe is expanding, this problem is resolved because the most distant galaxies will be red-shifted out of

_____

which end in a big crunch but, as far as we can tell, we live in a flat universe which is infinite in extent both in space and (in the forward direction) also in time sight.

Now the popular (but as we shall see, incorrect) picture is of a massive explosion which happened $T_0 = 13.8$ billion years ago when the universe was infinitely dense and when, in a sense, all the galaxies were in the same place, here, at the origin. Assuming for the moment that this is correct, and if we ignore the condition of special relativity that nothing can travel faster than light, it is easy to see that the relation between velocity and distance is simply

$$v = \frac{D_{real}}{T_0}$$

But this galaxy will not appear to be at a distance D_{real} because the light which it emits has to get back to us. Suppose that the light we can see was emitted when the galaxy was at a distance $D_{apparent}$ from us. The galaxy took a time $D_{apparent}/v$ to get to the point where it emitted the light and the light took $D_{apparent}/c$ (where c is the velocity of light) to get back to us. We therefore have

$$\frac{D_{apparent}}{v} + \frac{D_{apparent}}{c} = \frac{D_{real}}{v} = T_0$$

from which we can deduce that

$$D_{apparent} = \frac{D_{real}}{1 + D_{real}/c T_0}$$

What this implies is that galaxies which are a long way off appear closer than they really are.

We must also remember that since the more distant galaxies are receding faster, they will show a greater red shift and, in fact, any galaxy which is moving faster than light will be red shifted out of our vision.

In short, out universe will look something like this:

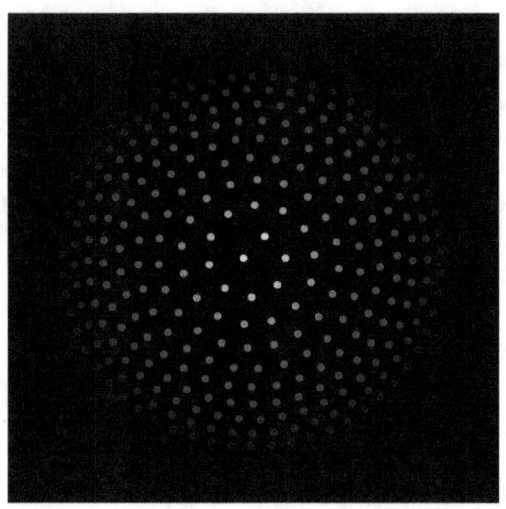

Distant galaxies would appear more numerous and closer together but they will appear to have the same angular size which you would expect them to have at that distance because the universe we are considering is flat (i.e. Euclidean). They will, however, be more and more red shifted and will eventually be red-shifted out of sight. The observable horizon will be at a distance of $\frac{1}{2} cT_0$ (i.e. about 7 billion light years) and even those galaxies at the extreme limits of our vision cannot have a value of D_{real} greater than 13.8 billion light years.

If we assume that the red shift is due to the standard formula that the guard and the station master worked out when they were discussing the colour of the signal (see page 66) we can show that

$$v = c \times \left(\frac{Z^2 - 1}{Z^2 + 1} \right) = \frac{D_{real}}{T_0}$$

A graph of $\frac{Z^2 - 1}{Z^2 + 1}$ against D_{real} should therefore be a nice straight line through the origin. The graph below shows a plot of the red shift Z of 94 galaxies plotted against their real distance[47] from us as measured

47 What I call the 'real distance' is more correctly known as the 'proper distance' and is the actual distance between us and the galaxy *at this time*. It is not possible to measure this distance directly but it can be inferred straightforwardly from other measurements in a way which is independent of the model used to describe the expansion of the universe. Unless otherwise stated, astronomers always mean 'proper distance' when they refer

by the Gamma Ray Burst technique[48].

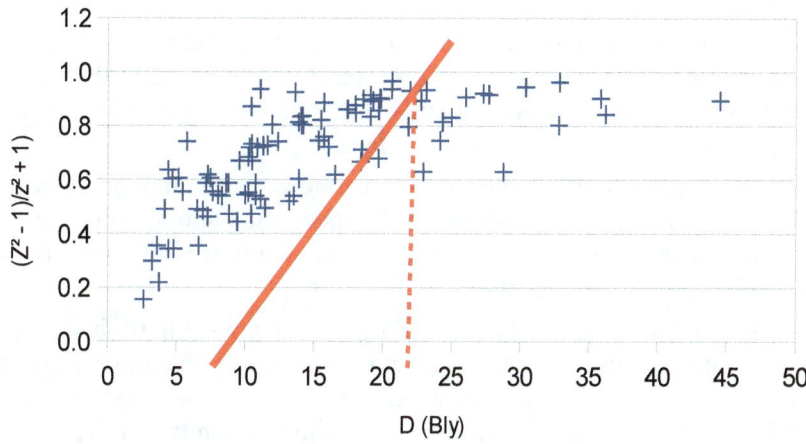

The red line on this and all the following graphs is the line which we would expect based on the observed recession of nearby galaxies – i.e. it passes through the point (13.8, 1)

Even making every allowance for the extreme uncertainty in measuring the distance to these extremely distant objects, it is clear that this simple picture of the way our universe is expanding is completely wrong. Not only is the trend line not straight, apparently we can see objects which have values of D_{real} which are way beyond 13.8 billion light years.

The flaw in the model is that assumption that the red shift of a distant galaxy is due to the speed of recession of the galaxy at the time the light is emitted. For nearby galaxies which are moving at much less than the speed of light, the approximation is a good one, but, when we are considering more distant galaxies, we must look for a new idea. The new idea is this: *the galaxies are not moving away from us at all*! What is actually happening is that *the scale which we use to measure distances is getting smaller.*

It is as if, on a certain day we measure the length of a race track and

---

to 'the distance to a galaxy'.

48 The data is taken from the NED-4D database which is available at
https://ned.ipac.caltech.edu/level5/NED0D/NED.4D.html

find it to be 100 m. A year we measure it again and we find that it is 110 m – not because the race track is longer than before but because our tape measure has shrunk.

An even better analogy is this. Suppose that one day a gold bar is valued at £10,000. A year later, however, it is valued at £11,000. The gold bar isn't really any more valuable than it was before – it just costs more because the value of the £ has fallen.

One image that is often used to illustrate an expanding universe is the image of a balloon being blown up. Spots on the balloon represent galaxies. The balloon expands but the spots themselves do not, only the space between them increases.

This picture is seriously misleading in two ways. Firstly it seems to suggest that the universe has to be curved. This is emphatically not the case. Our universe is (as far as we can tell) spatially *flat* and therefore probably infinite[49]. Secondly, it gives the impression that if the expansion rate was fast enough, an ant crawling round the balloon at a constant speed might be unable to reach a target spot because the spot would always be receding from it faster than it could approach. This is sometimes true, but not always, and not in the most important case as we shall see.

So to rid ourselves of any misconceptions about receding galaxies, let us agree on a new unit of distance, the *cosmological light year* or cly. We shall also agree that on the 1st of January 2000 CE (i.e. *now* in cosmological terms) 1 cly = 1 ly. A galaxy which is 1 Bly (billion light years) away from us now is also 1 Bcly away from us. Let us also agree to use the letter θ to represent distances in cosmological light years and D to measure normal distances.

Now the whole point about cosmological light years is that the galaxies *do not move* and distances measured in cosmological light years *never change*[50]. What changes is the *exchange rate* between

---

49 It is mathematically possible to conceive of a flat universe which is unbounded and yet finite but there is no reason to suppose that our universe is like this.

50 What I call the 'cosmological distance' is known to astronomers as the 'comoving distance'. By definition, the 'proper distance' to a galaxy *at this time* is equal to its 'comoving distance' but as time proceeds, the 'proper distance' increases while the 'comoving distance' stays the same.

cosmological light years and ordinary light years. We therefore have

$$D = a\theta$$

where a is the exchange rate between cosmological light years and ordinary light years.

Now, in general a is a function of time and in an expanding universe, it increases with time. The more time passes, the bigger a gets and the greater the measured distance D becomes between two galaxies separated by a fixed cosmological distance θ.

When a photon moves through this expanding space, every year it moves though 1 ordinary light year. But as a increases, the rate at which it moves through cosmological space (in cly per year) decreases. Mathematically we say that

$$\frac{d\theta}{dt} = \frac{c}{a}$$

where c is the velocity of light.

Basically what we are saying is that the rate at which a photon moves through cosmological space gets slower and slower. It also means that in the past, light moved through cosmological space faster than 1 cly per year. This does not violate Special Relativity because at all times, the light travels one ordinary light year in a year.

We now need to make an assumption about how a changes with time. The simplest assumption is that it increases linearly i.e. a is proportional to time t. At first sight it would appear that this assumption is exactly the same as the classical model which we analysed earlier because it implies that the rate at which the real distance between us and a certain galaxy is increasing is proportional to its distance from us – but this is not the same as saying that the galaxies are moving away from us at a speed which is proportional to the distance. In the classical universe, the galaxies were *actually moving* through fixed space (and violating the laws of special relativity too!). In this case, the galaxies are fixed; it is the *scale* of the universe which is changing.

To put it another way, in the classical expanding universe, at any given instant, the *speed of recession* is proportional to D_{real} (and since speeds cannot exceed the speed of light, galaxies with D_{real} greater than cT_0 do not exist); but in the relativistic universe, it is the *rate of increase* of D_{real} which is proportional to D_{real} and it is perfectly possible for galaxies with D_{real} greater than cT_0 to exist.

If you still think that this is just mincing with words, it is not; it makes a big difference in two ways. First it makes a difference to the length of time it takes for a photon to cross the gap between the distant galaxy and ourselves; and second, it makes a big difference to the observed red shift. Imagine a lorry load of soldiers travelling away from you at a constant speed of 20 mph. Every few seconds a soldier jumps out of the lorry and runs back towards you at 10 mph. Obviously the soldiers will take longer and longer to reach you because each successive soldier has a greater distance to run (this is the classic Doppler effect) but the soldiers will have no difficulty in reaching you because they are all running towards you at 10 mph. This is the classical view of an expanding universe.

Now, suppose instead that there is a huge rubber belt connecting you to the receding lorry and that the soldiers have to run back to you along the belt. At the instant one of the soldiers jumps out of the lorry onto the belt, even though he is running towards you at 10 mph, the belt is carrying him away from you at 20 mph so, initially at least, he will still be moving away from you. The question is – will he ever reach you?

The astonishing answer is – yes, he will eventually get to you regardless of the speed of the lorry! (But obviously it will take him a lot longer than it would have done in the classical case.) It is true that, initially the belt will be carrying him rapidly away from you but in time, as he makes his way down the belt, the speed of the expanding belt will reduce (remember the speed of the lorry is constant so the speed of the belt at any point along it will be a proportion of this speed) and eventually he will be able to make real progress towards you.

What we need is a mathematical formula relating the real distance which is travelled by a photon which sets out at a time T_1 and reaches its destination at a time T_2. But to get this we have to do a tiny bit of serious mathematics.

Since we have agreed that $a = 1$ when $t = T_0$ (where $T_0 = 13.8$ billion years) we have

$$a = \frac{t}{T_0}$$

hence

$$\frac{d\theta}{dt} = \frac{c}{a} = \frac{cT_0}{t}$$

In order to find the distance between the galaxies (in cosmological light years) we must integrate this expression from T_1 to T_2. The result is

$$\theta = \int_{T_1}^{T_2} \frac{cT_0}{t}\,dt = cT_0\log\frac{T_2}{T_1}$$

(where the logarithm is the *natural* logarithm to base e).

Now, what we are really interested in is the situation when the second galaxy is our own and when T_2 is *now* – i.e. when $T_2 = T_0$. We are not really very interested in the cosmological distance θ either; what we really want to know is the ordinary distance – i.e. the actual distance to the galaxy *now*. But remember – we have agreed to make the exchange rate unity *now*, so $D_{real} = \theta$.

Putting these facts into the last equation we arrive at the following important expression

$$\boxed{\theta = D_{real} = cT_0\log\frac{T_0}{T_1}} \tag{A}$$

This result has a number of surprising consequences. First the log function increases without limit as T_1 decreases. Therefore, as T_1 approaches zero (the big bang), D_{real} approaches infinity. Technically, therefore, in the linearly expanding universe, every galaxy that has ever existed is visible to us! To put it another way, here is no observable 'horizon' in this sort of universe.

Second, any galaxy which is sufficiently far away (so that $\log T_0/T_1$ >1) is now at a greater distance from us than the distance which light could have travelled in the whole age of the universe. The corollary of this is that, since all the galaxies were essentially in the same place when the universe began, the distance between us and such a galaxy must be increasing at a rate which is greater than the speed of light. This sounds absurd, but it is not. Remember the soldiers running at 10 mph away from a lorry travelling at 20 mph? Initially, they start by going backwards but, eventually, they reach that part of the belt which is only moving at less than 10 mph and then they can make more and more rapid progress towards their destination. Remember also that the galaxy is not *moving through space* at a speed which is greater than the speed of light, it is just that the space between us and the galaxy is increasing rapidly.

But this equation does not tell the whole story. It should be remembered that when we take a picture of the night sky, we are not seeing the galaxies as they are *now* but as they were *when the light left them*. Since the universe has expanded by a factor T_0/T_1 since then, the distance which we will actually see $D_{apparent}$ is less than D_{real} and given by:

$$D_{apparent} \ = \ D_{real}\frac{T_1}{T_0} \ = \ c\,T_1\log\frac{T_0}{T_1} \qquad\text{(B)}$$

This is a remarkable result and deserves some study. First let us remind ours~~~~~~e whose scale factor~~~~~~is is a fairly good appro:~~~~~~ possibilities~~~~~~irstly lets take a look

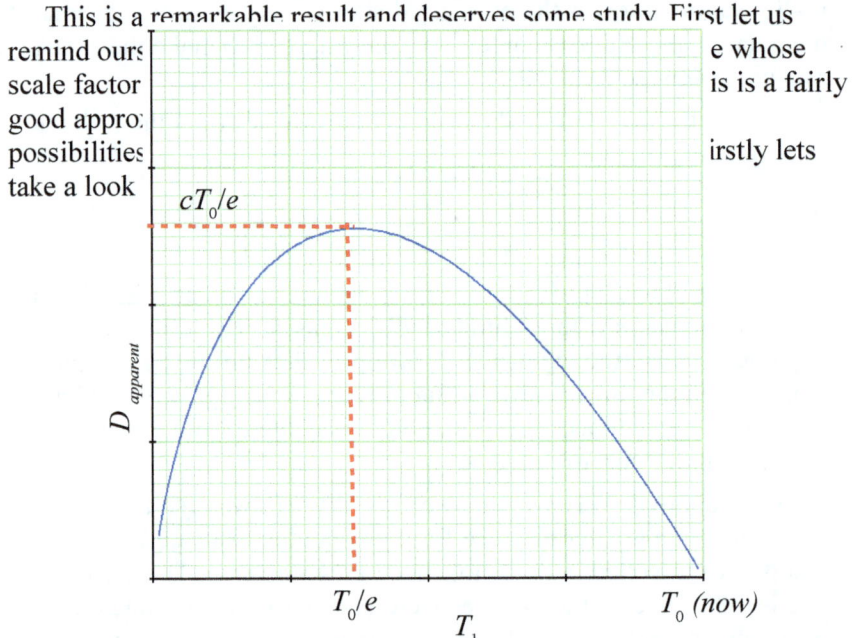

The graph shows the apparent distance of galaxies whose light we can see $D_{apparent}$ (i.e. all of them!) as a function of the times since the big bang at which the light left them T_1.

The weirdest thing about this graph is that it has a maximum. As we look further and further back in time, the galaxies apparently get further away and smaller, but galaxies whose light left them earlier than about 5 billion years after the big bang start to look closer and get bigger! The reason for this is simple. When the light left them they *were* closer and therefore looked a lot bigger! However, although they were close to us at that time, they were quite a long way away in cosmological terms, so it has taken a long time for their light to reach us.

In summary, there are three quantities of interest when we examine a galaxy given the time T_1 at which the light we can see left the galaxy: D_{real} (equation A). $D_{apparent}$ (equation B) and the angular size of the galaxy (assuming that all galaxies are the same size) which is proportional to $1/D_{apparent}$. Plotting all three quantities against T_1 gives us:

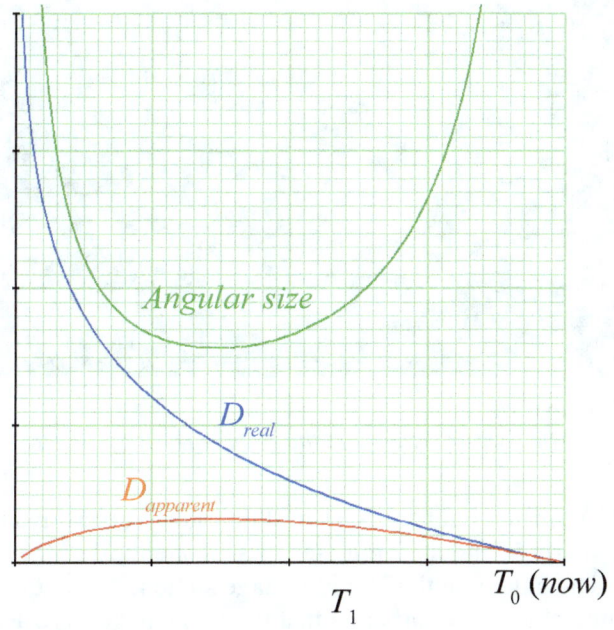

Take a look back at the Hubble Extreme Deep Field image on page 140. Halfway up on the left hand side there is a lovely spiral galaxy which looks very similar to the Whirlpool galaxy M51.

Whirlpool Galaxy M51
NASA, ESA, S. Beckwith (STScI) and the Hubble Heritage Team
(STScI/AURA)

Now, the Whirlpool galaxy is 31 million light years away and subtends an angle of about 10 minutes of arc (one third of the size of the full moon). The galaxy in the Hubble image subtends an angle of about 0.075 minutes of arc. If we assume that the two galaxies are about the same size, the galaxy in the image ought to be 130 times further away – i.e. about 4 billion light years away.

But this simple calculation is flawed because it does not take into account the expansion of the universe. It *looks* as if it is 4 billion light years away based on its angular size – but there are *two* possible solutions to the equation for T_1 given that $D_{apparent}$ = 4 billion light years.

From equation (B), and remembering that if we work in years and light years, $c = 1$), we have

$$4 = T_1 \log \frac{13.8}{T_1}$$

and the solutions to this are $T_1 = 8.5$ billion years and 2.2 billion years.

Substituting these figures into equation (A) we find that the galaxy could either be 6.7 billion light years or 25.3 billion light years away – it would look the same size in either case. Only by looking at its spectrum (particularly in the ultra violet region) can we tell which it is.

If it is 6.7 billion light years away, then it is receding from us at nearly 50% of the speed of light[51]. If it is actually 25.3 billion light years away, then it is receding from us at getting on for twice the speed of light[52] and light started out on its journey to us when the universe was only 16% [53]of its present age![54]

---

51 6.7 / 13.8 × 100 = 48.5%

52 25.3 / 13.8 × 100 = 183%

53 2.2 / 13.8 × 100 = 16%

54 Of course, these figures are illustrative only as they depend on the arbitrary assumption that the galaxy in the image is the same size as M51.

Cosmological Red Shift

The evidence for an expanding universe is simple and compelling. The further away a galaxy appears to be (on the basis of its brightness and angular size) the more its light is red-shifted. The red shift of a galaxy is measured by comparing the wavelength of a recognizable line in its spectrum with the wavelength of the same line as measured in the lab on Earth. If the measured wavelength is λ and the original wavelength is λ_0, then the red shift $Z = \lambda/\lambda_0$[55]. This number starts at 1 and increases without limit.

If we interpret the red shift, not as a Doppler shift, but as a measure of the amount by which the universe has expanded since the light was emitted from the galaxy, then we can simply put $Z = \dfrac{a(T_2)}{a(T_1)}$ where $a(T_1)$ and $a(T_2)$ are the scale factors of the universe at times T_1 and T_2.

In a linearly expanding universe, a is simply proportional to T so we can simplify this to $Z = \dfrac{T_2}{T_1}$ where T_1 is the time the light was emitted by the galaxy and T_2 is the time it arrived.

On page 149 we proved that in a linearly expanding universe the cosmological distance travelled by a photon between the T_1 and T_2 was

$$\theta = c T_0 \log \frac{T_2}{T_1} \qquad \text{(equation A)}$$

from which we can infer that

$$\theta = c T_0 \log Z$$

or

$$\boxed{Z = e^{\theta/cT_0}} \qquad \text{(C)}$$

In other words, in a linearly expanding universe, the red shift of a galaxy *depends only on its cosmological distance from us* and since that never changes, its red shift will not change either throughout the history of the universe.

---

55 Astronomers define red shift as $z = (\textit{increase in } \lambda)/\lambda$ which implies that $z = Z - 1$. When talking about cosmological distances, Z is preferred to z but it must be remembered that in all astronomy books, z will be quoted, not Z.

More specifically, since by definition, at this time $D_{real} = 0$, our expected relation between Z and D_{real} (equation (A) on page 149) simplifies to

$$D_{real} = cT_0 \log Z$$

and if our model of a linearly expanding universe is correct, a graph of the natural logarithm of Z against D_{real} ought to be a nice straight line through the origin.

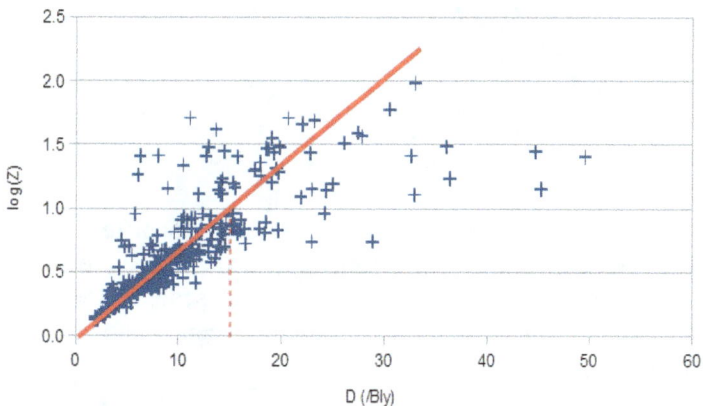

The reason for the huge scatter is, of course, the immense difficulty in measuring the distance to the most distance galaxies. There are many methods, the most accurate being the study of type 1 supernovae, but it has to be admitted that most practising scientists would be appalled at the prospect of deriving any valid conclusions from such a graph. Notwithstanding the size of the potential errors, however, the data are at least compatible with the idea that the universe has expanded linearly over time since about 14 billion years ago.

But this is not the only possibility. Einstein's theory predicts that such a linear relationship is actually only true of a completely empty universe! For a universe which contains a preponderance of matter, it turns out that a is proportional to $t^{2/3}$. In this case the relation between D and Z turns out to be:

$$D = 3cT_0\left(1 - \sqrt[3]{1/Z}\right)$$

The appropriate graph is shown below. It is clear that, on this model, the most distant galaxies consistently show a smaller red shift than we would expect.

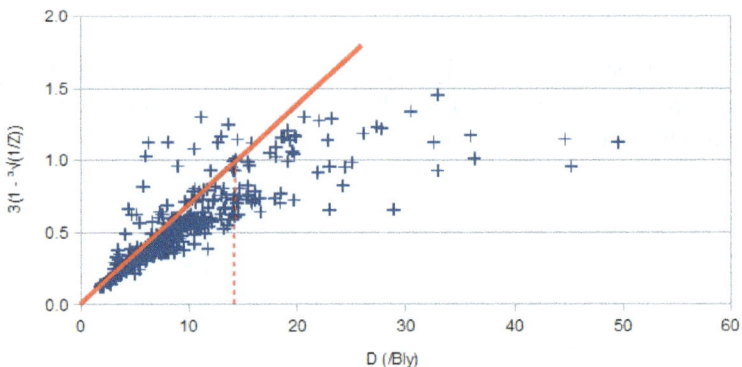

Finally, consider a third possibility which is consistent with Einstein's equations and which is characteristic of a universe which contains nothing but radiation. In this case the scale factor increases exponentially with time and the relation between D and Z turns out to be particularly simple:

$$D = cT_0(Z - 1)$$

The graph of $(1 - Z)$ against D is:

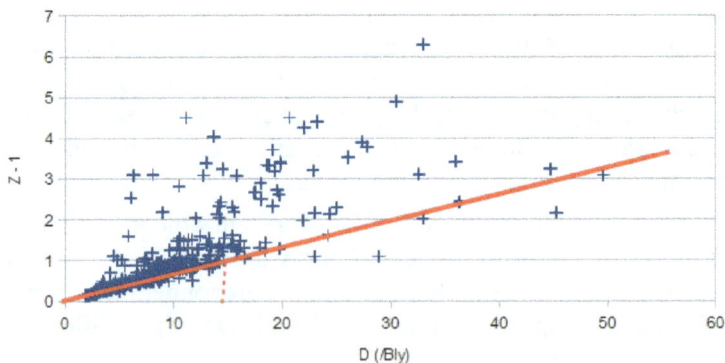

This graph shows exactly the opposite effect – distant galaxies showing a greater red shift than we would expect.

The basic problem is that none of these three models fits the data all the time, because the character of the universe has changed during the course of its evolution. In the early stages it was dominated by matter, which caused the initially very rapid expansion to slow down, but later it began to expand linearly and it is now believed to be expanding exponentially. This model is known as the ΛCDM model (pronounced 'lambda-C-D-M') which juggles with the various parameters in Einstein's Field Equations – notably the cosmological constant Λ which determines the amount of 'dark energy' in the universe and the amount of Cold Dark Matter in the universe – to get the best possible fit with the astronomical data. (One is inevitably reminded here of the way Ptolemy juggled with his epicycles to 'save the appearances' of the orbits of the planets!)

NASA has published the following image which is supposed to illustrate the history of the universe since the big bang, but it grossly over emphasises the inflationary phase of the expansion by using a disgracefully non-linear scale in both the temporal and spatial dimensions.

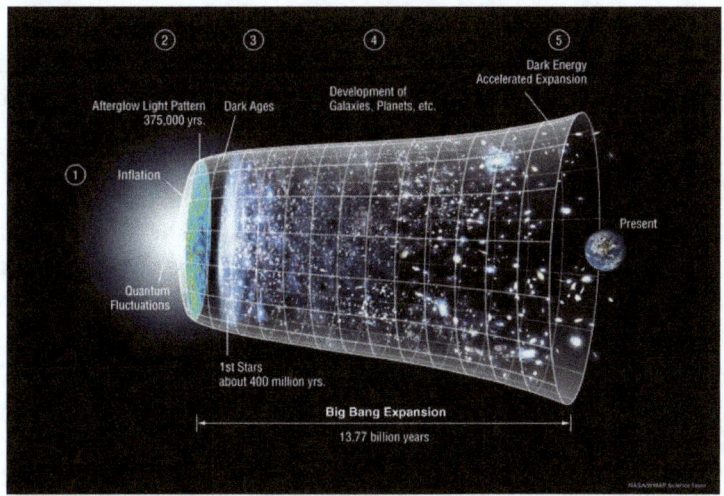

NASA/ LAMBDA Archive / WMAP Science Team

A much better illustration is to be found in Wikipedia which compares several different possible scenarios, all of which are compatible with Einstein's equations.

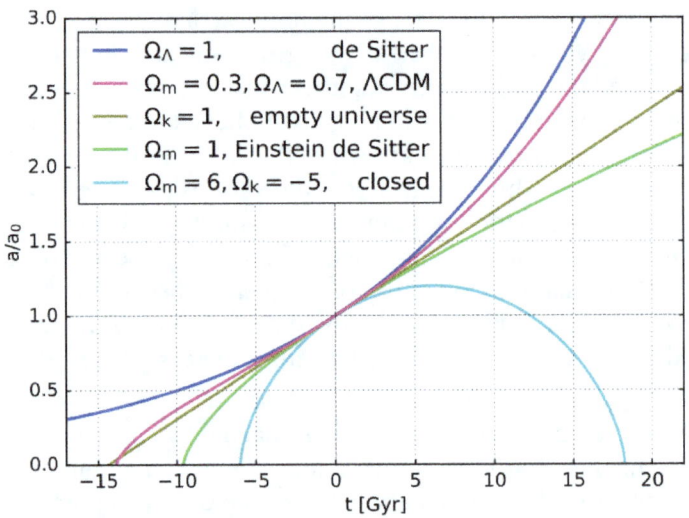

The de Sitter universes and the closed universe are of academic interest only. The ΛCDM universe (represented by the red line) begins by expanding exponentially for a very short while, then settles down to a period of expansion in which matter predominates, and now it appears to be entering a second phase of slow exponential expansion. By a sheer fluke, the age of the ΛCDM universe turns out to be almost exactly the same as the age of the linear universe (based on extrapolating back the apparent recessional speeds of nearby galaxies). It is for this reason, therefore, that to a good approximation, we shall not go far wrong if we assume that the expansion has been linear, that the age of the universe is about 13.8 billion years and that the relation between red shift and proper distance is $D_{real} = cT_0 \log Z$.

The red shift of the most distant galaxy measured to date (2021) (GN-z11) has a red shift of $Z = 12$ (i.e. $z = 11$) which, using the currently accepted value of T_0 as 13.8 billion years, puts its current distance (D_{real}) at 34 billion light years and its apparent distance $(D_{apparent})$ at 2.8 billion light years. The light which we are receiving from this galaxy started off on its journey 1.15 billion years after the big bang and its recessional speed is (and always has been) two and a half times the speed of light.

The Cosmic Microwave Background Radiation started off as radiation at about 3000K. Its temperature has now dropped to 2.7K – so it has been red shifted by a factor $Z = 1100$. This puts the proper distance (D_{real}) to the CMBR to about 97 billion light years. In practice this is the limit of the visible universe, because everything behind the CMBR is shrouded in a plasma fog. (But recently it has been suggested that we might be able to use gravitational waves to peer behind this curtain.)

It is sometimes suggested that during the inflationary period "*the universe expanded from much much smaller than a proton to about the size of a grapefruit — an increase in size of at least 10^{78}, or in plain English, a million times a million, and repeat 11 more times.*"

Quote taken from 'Dr Karl's Great Moments in Science' web page https://www.abc.net.au/science/articles/2013/08/06/3818354.htm

This is seriously misleading. As far as we know the universe is infinite – and always has been. It is true that the *scale* of the universe may have expanded by a factor of 10^{78}, but it never was the case that the universe was the size of a proton or a grapefruit, it was just incredibly dense.

There is, however, some sense in talking about how the *observable* universe has grown in size, but we must first define what we mean by the observable universe. As we have seen, in a linearly expanding universe, there is no limit to what is observable. In practice, however, it make sense to regard galaxies whose proper distance is less than cT_0 as being 'observable'. This volume is known as the Hubble volume or Hubble sphere. Obviously the Hubble sphere currently has a radius of 13.8 billion light years.

At the time of the decoupling event when the CMBR was created, the universe was 1100 times smaller than it is now and the Hubble sphere therefore had a radius of 12.5 million light years which is about the distance to one of our nearer galactic neighbours outside the local group M81.

If we consider the time when the Hubble sphere was the size of a grapefruit we are talking about a time after the big bang equal to the time it takes for light to travel about 6 cm, which computes to 0.2 ns.

I am afraid I can muster little enthusiasm for discussions about the behaviour of the universe at these extreme times. Once again, I am reminded of the intense debates which exercised the minds of medieval theologians concerning the number of angels who could dance on the head of a pin.

The Shrinking Universe

'Let me get this right' said Arthur as he and his sister were returning from a lecture on cosmology. 'Am I to understand that, if we accept that the universe has been expanding linearly from about 13.8 billion years ago, then galaxies with red shifts than 2.718... are receding from us faster than the speed of light?'

'Yes, that's absolutely right.' answered Betty. 'Not only that, these galaxies have always been receding from us faster than light, even when the light was emitted.'

'How is that possible?' asked Arthur.

'Initially, the light will be carried away from us by the expansion of the universe ,but, just like the soldiers on the rubber track[56], eventually the light will reach a point where the universe is not expanding so fast and it will start to make real progress.'

'Yes I sort of get that – but it seems that you are arguing round in circles. Surely, since nothing can travel faster than light, it follows that there can't be any galaxies whose Z is greater than 2.718..'

'There are two things to say about that. In the first place there are dozens of galaxies with $Z > 2.7$. But more importantly, it appears that you still haven't quite shaken off the idea that these galaxies are moving through space. They are not. Apart from small local movements, the galaxies are essentially stationary in cosmic space. It is only the distance between them which is changing.'

'So why do we go banging on about an *expanding* universe? Surely it would be better to talk about a *shrinking*... something?'

'I entirely agree.' said Betty. 'You are quite right. The universe isn't actually expanding. You could, with equal justification, say that it is everything in it which is shrinking.'

'Including atoms and stars?'

'Yes. But remember, all our rulers are shrinking too so the *measured* diameter of an atom or a star remains the same. Basically the only thing which measurably changes is the wavelength of a photon which spends an appreciable amount of time in transit.'

---

56 see page 148

'Perhaps the idea of an expanding universe is easier to accept after all,' said Arthur. 'But what I don't get is this; if the universe is expanding, why don't the atoms and stars in it expand too?'

'Let's get one thing straight. In so far as fundamental particles like protons and electrons have a size at all, their size is determined by the fundamental constants of nature, in particular the Planck length and the Planck time[57]. If these remain constant over time, then the size of these particles will remain the same. Atoms and galaxies, however, are complex systems bound together by electromagnetic and other forces. These forces are much stronger than the force which is causing the universe to expand and so these objects stay the same size.'

'Can you prove that?'

'I think so. Suppose you tie a mass m on the end of a long string of length D (where D is millions of light years long!). What you are asking is, what is the tension in the string due to the expansion of the universe. Right?'

'Yes. Go on.'

'Well, in a linearly expanding universe the relation between the cosmological distance θ and D is $D = \dfrac{t}{T_0}\theta$. But in this case D is held constant and it is θ which changes, so we must turn the equation round: $\theta = \dfrac{DT_0}{t}$.'

'So as time proceeds,' said Arthur, ' the cosmological distance decreases.'

'That's right. The rate at which the cosmological distance decreases is a measure of the speed with which the mass is moving through space towards us and we can calculate it by differentiating the above expression. This gives us $\dfrac{d\theta}{dt} = -\dfrac{DT_0}{t^2}$. Now what we really want is the acceleration, not the speed. To get this we need to differentiate

---

57 The Planck length is defined as $\sqrt{\dfrac{hG}{2\pi c^3}}$ and is equal to 1.6×10^{-35} m. The Planck time is defined as $\sqrt{\dfrac{hG}{2\pi c^5}}$ and is equal to 5.4×10^{-44} s.

again: $\dfrac{d^2\theta}{dt^2} = 2\dfrac{DT_0}{t^3}$, so the tension in the string at this moment when $t = T_0$ will be $2mD/T_0^2$.'

'Let's put some figures in and see what we get.' suggested Arthur.

'I don't think there is a lot of point in doing that' said Betty. 'Even if D is quite large, the T_0^2 term on the bottom is going to make the answer absolutely minute. What would be more interesting is to calculate how what length of string would make the cosmological force greater than, say, the force of gravity due to the Milky Way.'

'You mean by putting $\dfrac{2\,mD}{T_0^{\,2}} \geq \dfrac{GmM}{D^2}$ '

'That's right. The mass of the test mass m cancels leaving the following expression $D \geq \sqrt[3]{\dfrac{GMT_0^{\,2}}{2}}$ where M is the mass of the Milky Way.'

'That's not going to be easy to calculate. Give me a minute while I look up the figures and do the calculations.' said Arthur.

'Okay, I have done it.' he said a few minutes later. 'The answer comes to 2.8 million light years.'

'That's good.' said Betty. ' The Andromeda galaxy is part of our local group of galaxies and that is 2.5 million light years away. The nearest sizeable galaxy outside our local group is NGC500 which is about 6 million light years away – just outside the influence of the Milky Way.'

'That's fascinating. I think I am getting a much better picture of what it means to live in an expanding universe. But there is still something niggling at the back of my mind.'

'What's that?' asked Betty.

'I can see now why atoms and small groups of galaxies are not affected by the cosmological expansion, but I still don't see why photons expand.'

'That is a very good question.'

'Could it be that photons, sort of, get tired as they travel through space and lose energy that way, increasing their wavelength in the process?'

'No. I don't think so,' said Betty. 'You must remember that quantities like energy and wavelength only have meaning with respect to some observer. When an atom of hydrogen on a distant star emits a photon of light of wavelength 656.3 nm, this is the wavelength that would be measured by someone in the same locality and in the same inertial frame. When this photon is detected ages later when the universe has expanded by a factor of 2, its wavelength will be measured as 1312.6 nm, not because the photon is really any different but because the observer doing the measuring is not in the same inertial frame. Relative to the universe, his rulers have shrunk by a factor of 2 and so he sees the wavelength as being longer.'

'I don't buy that at all.' objected Arthur. ' You haven't really explained why the wavelength of the photon doesn't shrink (relative to the universe) like all the atoms in the observer's ruler. It seems that you are applying one rule to atoms and another to photons.'

'I admit you have a point there,' conceded Betty. 'I think the vital fact is that photons can be Doppler shifted in a way that atoms and other particles can't'

'What do you mean?' asked Arthur.

'Well, as you know, the standard relativistic formula for the Doppler shift of a photon emitted by a source receding from you at a speed v is

$$Z = \sqrt{\frac{c + v}{c - v}} \quad \text{[58]}$$

'Let's consider a photon emitted from a galaxy which is receding from us at the speed of light. According to the theory of a linearly expanding universe, $D_{rel} = cT_0$ and the expected Doppler shift ought to be $Z = 2.718...$ [59] But, of course, if we put $v = c$ than the whole expression blows up and we get an infinite answer.

'Now let's place a relay station halfway between the galaxy and ourselves[60]. Since, to an observer on the relay station, the galaxy is only a distance of $\frac{1}{2}cT_0$ away, the apparent recession velocity will be $\frac{1}{2}c$ and the measured Doppler shift will be $\sqrt{\frac{c + \frac{1}{2}c}{c - \frac{1}{2}c}} = \sqrt{3}$. This observer

---

58 See page 66
59 See equation C on page 154
60 Like the galaxy and Earth, the relay station must be 'comoving' – i.e. stationary, cosmologically speaking.

now passes the signal on to us who will see a further shift of $\sqrt{3}$ making a total shift of 3'

'I think I see what you are driving at' said Arthur excitedly. 'If you put two more relay stations in between the gaps making four stages, the Doppler shift will be $\left(\sqrt{\dfrac{c + \frac{1}{4}c}{c - \frac{1}{4}c}} \right)^4 = (\sqrt{5/3})^4 = 25/9 = 2.77...$.'

'That's right,' said Betty. 'In the limit you can put a whole string of (comoving) observers between us and the galaxy, you will find that the expected Doppler shift is 2.718... just like you predicted.'

'That's really cool.' said Arthur. 'So the cosmological Doppler shift is really just the familiar relativistic Doppler shift but applied not once but continuously throughout the whole journey! That's fantastic!'

'Yes. I think so too. I don't know the details but I think we can be sure that in whatever way the universe expands, applying the relativistic Doppler formula to a whole chain of comoving observers will result in a Doppler shift which is equivalent to the standard formula for cosmological red shift

$$ z = \frac{a(T_2)}{a(T_1)} \quad ,$$

'But hang on a minute,' interrupted Arthur, ' Earlier you said that photons can be Doppler shifted in a way that atoms and other particles can't. Why can't particles be Doppler shifted too?'

Yes. What I said was very misleading. Particles can indeed be Doppler shifted – but that doesn't make them get bigger in the sense of expanding like the universe. It does, however change their apparent wavelength just like a photon. If a distant galaxy were to emit a neutron, say, its wavelength (which in Quantum Theory is defined as h/p where p is the momentum of the particle) will be increased by a factor of $\dfrac{a(T_2)}{a(T_1)}$ but, of course, it won't get any *bigger.* Likewise, atoms and even lumps of matter will lose momentum (and energy) due to the cosmological red shift, but they won't *expand.* That is the sense in which photons are essentially different from particles.

'I seem to remember an earlier discussion concerning a terrorist firing bullets from the back of a flat bed truck[61]. Is it the same as that?' queried Arthur.

'Yes, it is basically the same effect. It doesn't matter whether we are talking about photons or neutron or bullets, exactly the same laws apply.'

'Suppose you put a photon in a 1 metre cube box with perfectly reflecting walls so that it bounced round inside for ever and ever without getting absorbed. Over aeons of time as the universe expanded, the box would remain the same size for the reasons you have explained – but would the wavelength of the photon increase?'

'Another very good question!'

'Well? What is the answer?'

'I don't think you are going to like the answer much.' said Betty.

'Why not?'

'Because the answer is that the wavelength will stay the same.'

'But that's ridiculous! We have spent the last half hour discussing why photons travelling freely through space increase their wavelength due to the expansion of the universe. Are you telling me that the universe doesn't expand inside an enclosed box?'

'No, I am not saying that.'

'Well you must be wrong then, because if the space inside the box expands like the rest of the universe, the wavelength of the photon must increase with it – as we have agreed.'

' Okay then, think of it like this. You agree that the box itself stays the same size. Right?'

'Right.'

'So the walls of the box are stationary. Right?'

'Right.'

'Wrong! You said yourself that the space inside the box expands along with the rest of the universe, so the *cosmological* distance between the wall decreases and from a cosmological point of view, the walls are moving inwards.'

'That's not fair! The measured distance between the walls remains the

---

61 See page 94

same so the walls must be stationary.'

'Not from the photon's point of view.'

'I suppose so. What of it?'

'It means that, while the photon is in transit, its wavelength increases because of the expansion of the universe, but when it hits the (cosmologically) moving walls it is blue shifted by precisely the right amount to restore the wavelength to its original value.'

'Well, that is diabolically clever, I must say!' exclaimed Arthur.

'Yes, I suppose it is – but really that is not the way you should think about the situation. From the point of view of an inertial observer inside the box, the walls remain 1 m apart the whole time, the photon simply bounces round off the stationary walls and its wavelength remains unchanged.'

'But if the walls of the box were, as it were, detached and allowed to drift apart with the expansion of the universe, then there would be no blue shift and, from the point of view of our inertial observer, the] wavelength of the photon would increase?'

'You've got it! Well done!'

'Well, it has been a bit of a mind-blowing journey but I think I have got a much better understanding of relativistic cosmology than I would have dreamed possible a few weeks ago.' said Arthur. 'Thank you.'

'My pleasure,' said his sister.

Conclusion

Of all the remarkable things about the universe perhaps the most remarkable is that we can see it at all.

Just suppose our planet was shrouded, like Venus, in thick cloud. Night and day, the seasons and the tides, all would be a complete mystery. It is debatable whether physics would ever have got off the ground. It is probable that our understanding of the nature of gravity would have been limited to an Aristotelian tendency for things to fall into their natural place.

Suppose that our planet was cloud free but space was filled with interstellar dust, so that we could only see the solar system and a faint glow from all the stars outside it. Isaac Newton would have had sufficient information to work out his famous laws and Einstein might have been able to formulate his theories of Special and General Relativity – but would we ever have deduced that the Sun was not the centre of the universe, but was one of many stars which were part of a vast whirlpool we call our galaxy?

And if we could only see the stars in our galaxy but not the myriad of other objects that populate our universe outside the Milky Way, would we ever have realised that the universe was expanding and began a finite time ago?

Even granted the fact that our universe is remarkably transparent to photons, I still find it hard to believe that the photons which started out on their journey towards the Hubble Space Telescope from some unimaginably distant galaxy, and which have travelled for perhaps 10 billion years without being either absorbed by a particle of dust or deflected by a fraction of an arc-second by passing close to a lump of dark matter, can still be reconstructed to form a recognizable shape on a photographic plate.

So when next you gaze on the wonder that is the night sky or when you open a book of photos taken from the HST, you are bound to feel a sense of profound gratitude as well as awe; for if the universe had been ever so slightly different, we might have had no knowledge of it at all. As it is, advances in observational technologies and theoretical astronomy over the last three decades have given us an unprecedented amount of information about our universe but, as always, this information has often served to raise more questions than provide answers. As far as we can tell, on a sufficiently large scale the universe we live in is isotropic (i.e. it looks the same in all directions) and is homogeneous (i.e. is basically the same everywhere). But is it infinite or finite? Is its expansion accelerating or decelerating? Is it dominated by matter or dark energy? Is there enough matter in it to stop the expansion or will it expand for ever? How much of it can we see? No one knows for sure.

Einstein's theory of General Relativity has withstood every test that has been thrown at it for nearly a century and although there have been many attempts to modify it, it is still the best we have got at the moment. But Einstein's equations allow of too many possible universes and none of the possible candidates have that elegant simplicity which we crave.

There is another problem too.

We have another well attested theory which has been incredibly successful at explaining the behaviour of matter on a small scale – Quantum Theory. Normally the two theories apply to very different domains and do not come into conflict – but this is not the case when we need to talk about either black holes or the very early history of the universe. The trouble is that the two theories are based on ideas which are so far removed from each other, it is almost impossible to see how they can be reconciled. Einstein himself spent (wasted?) 40 years of his life trying to find what we are now pleased call the 'Quantum theory of Gravity' and many other great minds have joined the search without success. Most have taken the view that Einstein's theory is the one that is at fault – hence the search for a 'Quantum Theory of Gravity' rather than a 'Relativistic Theory of Quanta' – but I think this attitude may be mistaken. Quantum Theory differs from Einstein's theory in one interesting respect. Whereas the latter was dreamed up by a lone intellect working largely from *a priori* principles, the former was developed by many scientists arguing over details and methods of approach. The result is that while General Relativity stands or falls on a single basic principle – the principle of Equivalence – Quantum Theory has no such coherence. Indeed, the paradoxes which are generated by QT (the wave-particle duality, the collapse of the wave function, entanglement etc.) are of a different class than the paradoxes which we have been discussing in this book. They cannot be explained simply by some clear thinking and a bit of maths – they have a philosophical dimension which makes them impossible to resolve.

So which is ultimately going to give? General Relativity or Quantum |Theory?

So far in the history of science, *a priori* theories have either been incredibly successful or disastrously wrong. There are at least five examples of the former. Copernicus was not really trying to 'save the appearances' when he proposed his heliocentric system of the world – he just knew, deep down, that the whole system of planets made so much more sense if it was assumed that the Sun, not the Earth, was at the centre. Galileo could not prove experimentally that motion was a state, not something which had to be maintained, he just knew, deep down, that it was so and that, in the absence of any friction, things would go on moving for ever. Newton had no idea why inertial mass and gravitational mass cancelled out causing all things to fall with the same acceleration, he just knew, deep down, that it had to be like that. Maxwell knew, deep down, that his equations had to be symmetrical and Einstein knew, deep down, that somehow the laws of physics had to be the same in any frame of reference.

These principles are so fundamental that it seems to me very unlikely that any new successful 'Theory of Everything' could possibly deny any of them – but I could be very wrong. My confidence in *a priori* principles was severely shaken with the discovery that left-right parity was violated in some nuclear interactions, so we must try to keep an open mind.

We seem to be facing another crisis as well. In 1960 the physicist Eugene Wigner noted the 'unreasonable effectiveness of Mathematics' in describing the physical world. In fact, by 1960, it was already clear that Mathematics alone could not describe all aspects of the physical world, because Quantum Theory only predicts the *probability* of events, not the events themselves. No amount of maths will enable you to predict exactly where a photon will land or when a Uranium atom will decay.

It is not sufficient to argue that probability theory is just another branch of Mathematics. It is true that it is possible to prove mathematically that the probability of a number chosen *at random* in the region of 10^{43} being prime is about 1%, but it is not the case that the number $10^{43} + 1$ has a probability of 1% of being prime or, indeed that it is simultaneously 1% prime and 99% not prime. Either it is prime or it is not. We just don't know. The physical world, on the other hand, appears to obey the laws of chance, not the laws of logic, and is therefore fundamentally not mathematical.

Another famous quotation, this time from Einstein himself in 1936 goes as follows: "The most incomprehensible thing about the universe is that it is comprehensible." But there are several features of the ΛCDM model of our universe that remain stubbornly incomprehensible. I refer to the problems of the so-called 'dark matter' and 'dark energy'. Now, it may very well be that these issues will soon be resolved, but I cannot help wondering if we are nearing the end of the road in our attempts to comprehend the universe we live in on both large and small scales. It is not as if we actually need to discover any new fundamental particles in order to feed the world's population or that we are actually ever going to be able to harness the energy of a rotating black hole to provide us with the energy we need. We have quite enough knowledge already to provide for the needs of the world – all that is lacking is the political will to implement the solutions.

On the other hand – who knows? Perhaps some future genius thinking in his ivory tower will come up with a new fundamental *a priori* principle which will miraculously unify and explain everything and set Physics and Astronomy back on the road to a deeper comprehension of our wonderful universe.

Any suggestions?

Appendix

The classic Doppler effect in sound

There are two Doppler effects – the Moving Source effect and the Moving Observer effect. The former is most often associated with the sound of the siren of an ambulance or police car as it goes by; the latter is heard on the continent when a train passes a level crossing sounding a bell. Although the two effects sound very similar, they are, in fact, quite different.

First, the Moving Source effect. As the ambulance moves forwards, the sound waves in front of the ambulance are bunched up because the ambulance is chasing the sound. If the ambulance is moving at a speed v and the sound is moving (through the stationary air) at a speed c, the sound waves are going to be reduced in length by a factor $\dfrac{c - v}{c}$. (You can check that this is the right factor because if $v = 0$, the factor equals 1; and if $v = c$, the factor equals zero).

Now, if the wavelength is reduced by a factor $\dfrac{c - v}{c}$ the frequency of the sound will be increased by the factor $\dfrac{c}{c - v}$. Now an increase in frequency of 6% causes the pitch of a note to go up by one semitone. Since the speed of sound in air is about 740 mph, the siren of an ambulance travelling towards you at 6% of this i.e. 45 mph, will cause the pitch to rise by one semitone. Similarly, as it travels away from you the pitch will drop by one semitone making the total difference one whole tone.

It is worth noting that if the object moves towards you at the speed of sound, the frequency becomes infinite – in other words you hear all the sound at once. This is essentially the cause of a sonic boom caused by an aircraft travelling faster then sound.

The Moving Observer effect is different because it affects the frequency of the sound directly, not its wavelength. If you are approaching a source of sound at a speed v the frequency of the waves as you cross them is going to increase by a factor $\frac{c + v}{c}$ and the wavelength shortened by $\frac{c}{c + v}$. If v is much smaller than c then this formula is not very different from the Moving Source formula but if $v = c$, the frequency is simply doubled – there is no sonic boom.

On the other hand, if you are moving *away* from the source of sound, v is negative and if you move fast enough you can make the frequency equal to zero, in which case you won't hear the sound at all.

The γ factor equation

$$\gamma_{u+v} = \frac{1}{\sqrt{1 - \left(\dfrac{c(u + v)}{c^2 + uv}\right)^2}}$$

$$= \frac{c^2 + uv}{\sqrt{(c^2 + uv)^2 - c^2(u + v)^2}}$$

$$= \frac{c^2 + uv}{\sqrt{c^4 + 2c^2uv + u^2v^2 - c^2u^2 - 2c^2uv - c^2v^2}}$$

$$= \frac{c^2 + uv}{\sqrt{c^4 - c^2u^2 - c^2v^2 + u^2v^2}}$$

$$= \frac{c^2 + uv}{\sqrt{(c^2 - u^2)(c^2 - v^2)}}$$

$$= \frac{1 + uv/c^2}{\sqrt{(1 - u^2/c^2)}\,\sqrt{(1 - v^2/c^2)}}$$

$$= \gamma_u \gamma_v(1 + uv/c^2)$$

The apparent speed of a passing space ship

The apparent speed v' of a space ship moving from left to right at a speed v at a perpendicular distance D is $v' = v\dfrac{c}{c + v\sin\theta}$ where

$\sin\theta = \dfrac{x}{\sqrt{x^2 + D^2}}$ (see page 74).

To calculate the apparent distance travelled by the space ship (in other words, to calculate where the space ship will appear to be at every instant in time) we must do some integration. The time taken T for the object to reach a point X will be:

$$T = \int_0^X \frac{1}{v'}\, dx = \int_0^X \frac{c + v\sin\theta}{cv}\, dx$$

Now $\tan\theta = \dfrac{x}{D}$ so $dx = D\sec^2\theta\, d\theta$

so

$$T = \frac{1}{v}\int dx + \frac{D}{c}\int \sin\theta\sec^2\theta\, d\theta$$

Now fortunately, $\sin\theta \sec^2\theta\, d\theta$ is a standard integral and is equal to $\sec\theta$ so

$$T = \frac{X}{v} + \frac{D}{c}\left(\sec\tan^{-1}\frac{X}{D}\right) + C$$

$$T = \frac{X}{v} + \frac{1}{c}\left(\sqrt{X^2 + D^2}\right) + C$$

Now when $X = 0$, T will not be equal to 0 because light from the object takes time to reach the observer so T will be equal to D/c. It follows therefore that the constant of integration $C = 0$

hence

$$T = \frac{X}{v} + \frac{1}{c}\left(\sqrt{X^2 + D^2}\right)$$

The Time Dilation factor deduced from Betty's lift experiment

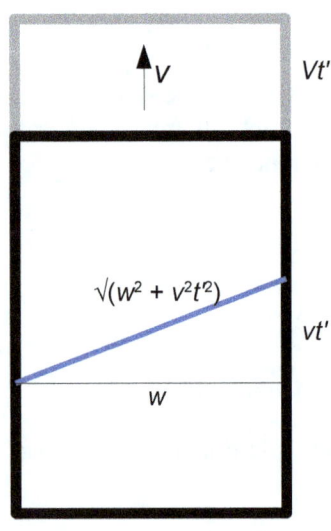

To an outside observer, the lift moves a distance vt' in a time t'. The light beam, however, moves a distance $\sqrt{w^2 + v^2 t'^2}$. Hence we have:

$$t' = \frac{\sqrt{w^2 + v^2 t'^2}}{c}$$

$$c^2 t'^2 = w^2 + v^2 t'^2$$

$$t'^2 = \frac{w^2}{c^2 - v^2} = \frac{(w/c)^2}{1 - v^2/c^2}$$

$$t' = \frac{w/c}{\sqrt{1 - v^2/c^2}}$$

To Betty inside the lift, she sees the light beam travel straight across the lift, a distance w in a time t. Hence

$$t = w/c$$

Eliminating w we get

$$t' = \frac{t}{\sqrt{1 - v^2/c^2}}$$

176

Gravitational Potential near a star

Provided that the gravitational field is small enough so that relativistic effects are negligible, the gravitational potential difference between a point on the surface and at a height h above a planet or star is given by the expression

$$\Delta\phi = \int_{R}^{R+h} \frac{GM}{r^2} dr = -GM\left[\frac{1}{r}\right]_{R}^{R+h} = \frac{GMh}{R(R+h)}$$

If the acceleration due to gravity at the surface of the planet is g_s then

$$g_s = \frac{GM}{R^2}$$

from which we get $\quad \Delta\phi = \dfrac{Rh}{R+h} g_s$

If the upper limit of the integral is ∞ then the equation becomes

$$\Delta\phi = \int_{R}^{\infty} \frac{GM}{r^2} dr = \frac{GM}{R} = Rg_s$$

It follows that the time dilation factor on the surface of a planet or star (relative to a distant observer is equal to

$$\frac{1}{\sqrt{1 - 2\Delta\phi/c^2}} = \frac{1}{\sqrt{1 - 2GM/Rc^2}} = \frac{1}{\sqrt{1 - 2Rg_s/c^2}}$$

(It is interesting to note that if $2GM/Rc^2 = 1$ then the time dilation factor becomes infinite. Turning this equation round gives us

$R = \dfrac{2GM}{c^2}$ which is actually the correct formula for the

Schwartzchild radius of the star. Little if any significance should be read into this, as the formula for the gravitational potential of the star is only valid in non-relativistic circumstances – a condition which is clearly not met near the event horizon of a black hole.)

Escape Velocity

On the previous page we showed that the gravitational potential at a height h above the surface of a star or planet is

$$\Delta \phi = \frac{GMh}{R(R + h)}$$

If we let h tend to infinity, the potential at infinity is

$$\phi_\infty = \frac{GM}{R}$$

and the energy needed to lift a mass m to infinity is

$$E = m\phi_\infty = \frac{GMm}{R}$$

If a projectile is launched from the surface with a velocity equal to the escape velocity then we have

$$\frac{1}{2}m v_{esc}^2 = \frac{GMm}{R}$$

$$v_{esc} = \sqrt{\frac{2GM}{R}}$$

(Of course, if we put $v_{esc} = c$ and turn the equation round, we get the correct expression for the Schwartzchild Radius of the object. As with the derivation on the previous page, little, if any, significance should be read into this.)

The Schwartzchild Radius

We have referred to the following formula for the Schwartzchild radius of a black hole on a couple of occasions (pages 107 and 131).

$$r_{sch} = \frac{2GM}{c^2}$$

Unfortunately there is no simple way to derive this formula without using a full mathematical treatment of the four dimensional space time around the object. Essentially it turns out that the interval ds along a radial line from a non-rotating black hole must be calculated using the (simplified) expression

$$ds^2 = \left(1 - \frac{2GM}{rc^2}\right)dt^2 - \left(1 - \frac{2GM}{rc^2}\right)^{-1}dr^2/c^2$$

where r is the 'circumferential radius' (i.e. the circumference of a circle round the black hole divided by 2π)

When r is very large, this expression reduces to the familiar expression for interval defined on page 57

$$ds^2 = dt^2 - dr^2/c^2$$

or
$$I = \sqrt{dt^2 - dx^2/c^2}$$

It is clear, however, that something strange is going to happen when

$$\frac{2GM}{rc^2} = 1 \quad.$$

The temporal term goes to zero and the spatial term blows up to infinity. It was this behaviour that confused everyone at first (including Einstein) into thinking that the event horizon was a singularity. As we now realize, it is only a singularity from the point of view of a distant observer. It is what is known as a coordinate singularity.